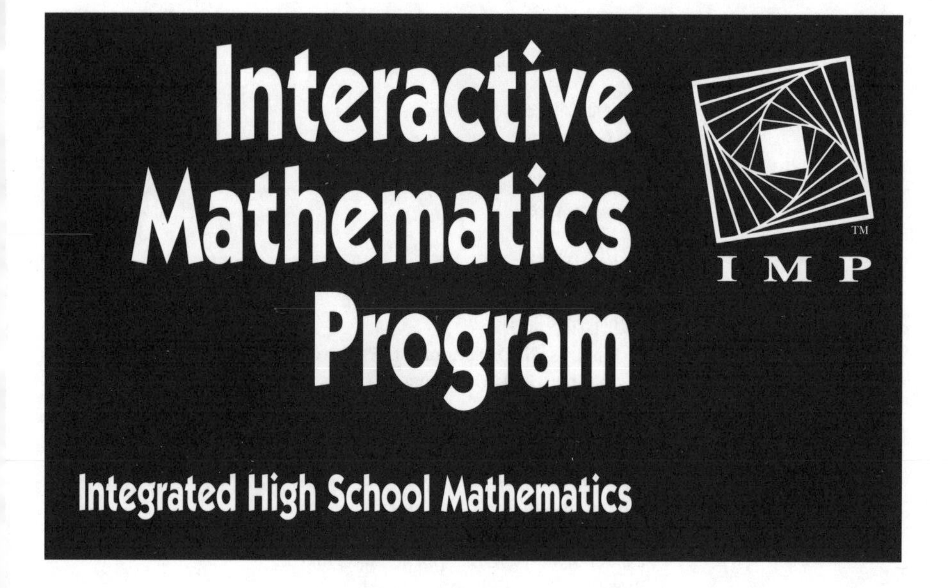

Cookies

Dan Fendel and Diane Resek
with
Lynne Alper and Sherry Fraser

KEY CURRICULUM PRESS
Innovators in Mathematics Education

This material is based upon work supported by the National Science Foundation under award number ESI-9255262. Any opinions, findings, and conclusions or recommendations expressed in this publication are those of the authors and do not necessarily reflect the views of the National Science Foundation.

Key Curriculum Press
1150 65th Street
Emeryville, California 94608
510-595-7000
editorial@keypress.com
http://www.keypress.com

10 9 8 7 6 5 4 3 04 03 02 01
ISBN 1-55953-261-0

Printed in the United States of America

Project Editor
Casey FitzSimons

Editorial Assistant
Jeff Gammon

Production Editor
Caroline Ayres

Art Developer
Jason Luz

Mathematics Review
Rick Marks, Ph.D., Sonoma State University
Rohnert Park, California

Teacher Reviews
Daniel R. Bennett, Kualapuu, Hawaii
Larry Biggers, San Antonio, Texas
Dave Calhoun, Fresno, California
Dwight Fuller, Clovis, California
Daniel S. Johnson, Campbell, California
Brent McClain, Hillsboro, Oregon
Amy C. Roszak, Roseburg, Oregon
Carmen C. Rubino, Lakewood, Colorado
Jean Stilwell, Minneapolis, Minnesota
Wendy Tokumine, Honolulu, Hawaii

Multicultural Reviews
Mary Barnes, M.Sc., University of Melbourne,
Cremorne, New South Wales, Australia
Edward D. Castillo, Ph.D., Sonoma State University,
Rohnert Park, California
Joyla Gregory, B.A., College Preparatory School,
Oakland, California
Genevieve Lau, Ph.D., Skyline College,
San Bruno, California
Beatrice Lumpkin, M.S., Malcolm X College (retired),
Chicago, Illinois
Arthur Ramirez, Ph.D., Sonoma State University,
Rohnert Park, California

Cover and Interior Design
Terry Lockman
Lumina Designworks

Production Manager
Steve Rogers, Luis Shein

Production Coordination
Diana Krevsky, Susan Parini

Technical Graphics
Kristen Garneau, Natalie Hill, Greg Reeves

Illustration
Tom Fowler, Evangelia Philippidis, Sara Swan,
Diane Varner, Martha Weston, April Goodman Willy

Publisher
Steven Rasmussen

Editorial Director
John Bergez

Acknowledgments

Many people have contributed to the development of the IMP curriculum, including the hundreds of teachers and many thousands of students who used preliminary versions of the materials. Of course, there is no way to thank all of them individually, but the IMP directors want to give some special acknowledgments.

We want to give extraordinary thanks to the following people who played unique roles in the development of the curriculum.

- **Matt Bremer** did the initial revision of every unit after its pilot testing. Each unit of the curriculum also underwent extensive focus group reexamination after being taught for several years, and Matt did the rewrite of many units following the focus groups. He has read every word of everyone else's revisions as well and has contributed tremendous insight through his understanding of high school students and the high school classroom.
- **Mary Jo Cittadino** became a high school student once again during the piloting of the curriculum, attending class daily and doing all the class activities, homework, and POWs. Because of this experience, her contributions to focus groups had a unique perspective. This is a good place to thank her also for her contributions to IMP as Network Coordinator for California. In that capacity, she has visited many IMP classrooms and has answered thousands of questions from parents, teachers, and administrators.
- **Lori Green** left the classroom as a regular teacher after the 1989-90 school year and became a traveling resource for IMP classroom teachers. In that role, she has seen more classes using the curriculum than we can count. She has compiled many of the insights from her classroom observations in the *Teaching Handbook for the Interactive Mathematics Program*™.
- **Bill Finzer** was one of the original directors of IMP before going on to different pastures. Though he was not involved in the writing of Year 2, his ideas about curriculum are visible throughout the program.
- **Celia Stevenson** developed the charming and witty graphics that graced the prepublication versions of all the IMP units.

Several people played particular roles in the development of this unit, *Cookies:*

- Matt Bremer, Janice Bussey, Donna Gaarder, Theresa Hernandez-Heinz, Linda Schroers, and Adrienne Yank helped us create the version of *Cookies* that was pilot tested during 1990-91. They not only taught the unit in their classrooms that year, but also read and commented on early drafts, tested almost all the activities during workshops that preceded the teaching, and then came back after teaching the unit with insights that contributed to the initial revision.

- Donna Gaarder, Diana Herrington, and Dan Johnson joined Matt Bremer, Mary Jo Cittadino, and Lori Green for the focus group on *Cookies* in June, 1995. Their contributions built on several years of IMP teaching, including at least two years teaching this unit, and their work led to the development of the last field-test version of the unit.
- Dan Branham, Dave Calhoun, Steve Hansen, Gwennyth Trice, and Julie Walker field tested the post-focus-group version of *Cookies* during 1995–96. Dave and Gwennyth met with us to share their experiences when the teaching of the unit was finished. Their feedback helped shape the final version that now appears.

In creating this program, we needed help in many areas other than writing curriculum and giving support to teachers.

The National Science Foundation (NSF) has been the primary sponsor of the Interactive Mathematics Program. We want to thank NSF for its ongoing support, and we especially want to extend our personal thanks to Dr. Margaret Cozzens, Director of NSF's Division of Elementary, Secondary, and Informal Education for her encouragement and her faith in our efforts.

We also want to acknowledge here the initial support for curriculum development from the California Postsecondary Education Commission and the San Francisco Foundation, and the major support for dissemination from the Noyce Foundation and the David and Lucile Packard Foundation.

Keeping all of our work going required the help of a first-rate office staff. This group of talented and hard-working individuals worked tirelessly on many tasks, such as sending out units, keeping the books balanced, helping us get our message out to the public, and handling communications with schools, teachers, and administrators. We greatly appreciate their dedication.

- Barbara Ford—Secretary
- Tony Gillies—Project Manager
- Marianne Smith—Communications Manager
- Linda Witnov—Outreach Coordinator

We want to thank Dr. Norman Webb of the Wisconsin Center for Education Research for his leadership in our evaluation program, and our Evaluation Advisory Board, whose expertise was so valuable in that aspect of our work.

- David Clarke, University of Melbourne
- Robert Davis, Rutgers University
- George Hein, Lesley College
- Mark St. John, Inverness Research Associates

IMP National Advisory Board

We have been further supported in this work by our National Advisory Board—a group of very busy people who found time in their schedules to give us more than a piece of their minds every year. We thank them for their ideas and their forthrightness.

David Blackwell
Professor of Mathematics and Statistics
University of California, Berkeley

Constance Clayton
Professor of Pediatrics
Chief, Division of Community Health Care
Medical College of Pennsylvania

Tom Ferrio
Manager, Professional Calculators
Texas Instruments

Andrew M. Gleason
Hollis Professor of Mathematics and Natural Philosophy
Department of Mathematics
Harvard University

Milton Gordon
President and Professor of Mathematics
California State University, Fullerton

Shirley Hill
Curator's Professor of Education and Mathematics
School of Education
University of Missouri

Steven Leinwand
Mathematics Consultant
Connecticut Department of Education

Art McArdle
Northern California Surveyors Apprentice Committee

Diane Ravitch (1994 only)
Senior Research Scholar
Brookings Institution

Roy Romer (1992-1994 only)
Governor
State of Colorado

Karen Sheingold
Research Director
Educational Testing Service

Theodore R. Sizer
Chairman
Coalition of Essential Schools

Gary D. Watts
Educational Consultant

Finally, we want to thank Steve Rasmussen, President of Key Curriculum Press, Casey FitzSimons, Key's Project Editor for the IMP curriculum, and the many others at Key whose work turned our ideas and words into published form.

Dan Fendel Diane Resek Lynne Alper Sherry Fraser

Foreword

Is There Really a Difference? asks the title of one Year 2 unit of the Interactive Mathematics Program (IMP).

"You bet there is!" As Superintendent of Schools, I have found that IMP students in our District have more fun, are well prepared for our State Testing Program in the tenth grade, and are a more representative mix of the different groups in our geographical area than students in other pre-college math classes. Over the last few years, IMP has become an important example of curriculum reform in both our math and science programs.

When we decided in 1992 to pilot the Interactive Mathematics Program, we were excited about its modern approach to restructuring the traditional high school math sequence of courses and topics and its applied use of significant technology. We hoped that IMP would not only revitalize the pre-college math program, but also extend it to virtually all ninth-grade students. At the same time, we had a few concerns about whether IMP students would acquire all of the traditional course skills in algebra, geometry, and trigonometry.

Within the first year, the program proved successful and we were exceptionally pleased with the students' positive reaction and performance, the feedback from parents, and the enthusiasm of teachers. Our first group of IMP students, who graduated in June, 1996, scored as well on PSATs, SATs, and State tests as a comparable group of students in the traditional program did, and subsequent IMP groups are doing the same. In addition, the students have become our most enthusiastic and effective IMP promoters when visiting middle school classes to describe math course options to incoming ninth graders. One student commented, "IMP is the most fun math class I've ever had." Another said, "IMP makes you work hard, but you don't even notice it."

In our first pilot year, we found that the IMP course reached a broader range of students than the traditional Algebra 1 course did. It worked wonderfully not only for honors students, but for other students who would not have begun algebra study until tenth grade or later. The most successful students were those who became intrigued with exciting applications, enjoyed working in a group, and were willing to tackle the hard work of thinking seriously about math on a daily basis.

IMP Year 2 places the graphing calculator and computer in central positions early in the math curriculum. Students thrive on the regular group collaboration and grow in self-confidence and skill as they present their ideas to a large group. Most importantly, not only do students learn the symbolic and graphing applications of elementary algebra, the statistics of *Is There Really a Difference?,* and the geometry of *Do Bees Build It Best?,* but the concepts have meaning to them.

I wish you well as you continue your IMP path for a second year. I am confident that students and teachers using Year 2 will enjoy mathematics more than ever as they experiment, investigate, and discover solutions to the problems and activities presented this year.

Reginald Mayo
Superintendent
New Haven Public Schools
New Haven, Connecticut

The Interactive Mathematics Program

What is the Interactive Mathematics Program?

The Interactive Mathematics Program (IMP) is a growing collaboration of mathematicians, teacher-educators, and teachers who have been working together since 1989 on both curriculum development and professional development for teachers.

What is the IMP curriculum?

IMP has created a four-year program of problem-based mathematics that replaces the traditional Algebra I–Geometry–Algebra II/Trigonometry–Precalculus sequence and that is designed to exemplify the curriculum reform called for in the *Curriculum and Evaluation Standards* of the National Council of Teachers of Mathematics (NCTM).

The IMP curriculum integrates traditional material with additional topics recommended by the NCTM *Standards,* such as statistics, probability, curve fitting, and matrix algebra. Although every IMP unit has a specific mathematical focus, most units are structured around a central problem and bring in other topics as needed to solve that problem, rather than narrowly restricting the mathematical content. Ideas that are developed in one unit are generally revisited and deepened in one or more later units.

For which students is the IMP curriculum intended?

The IMP curriculum is for all students. One of IMP's goals is to make the learning of a core mathematics curriculum accessible to everyone. Toward that end, we have designed the program for use with heterogeneous classes. We provide you with a varied collection of supplemental problems to give you the flexibility to meet individual student needs.

Teacher Phyllis Quick confers with a group of students.

How is the IMP classroom different?

When you use the IMP curriculum, your role changes from "imparter of knowledge" to observer and facilitator. You ask challenging questions. You do not give all the answers; rather, you prod students to do their own thinking, to make generalizations, and to go beyond the immediate problem by asking themselves "What if?" The IMP curriculum gives students many opportunities to write about their mathematical thinking, to reflect on what they have done, and to make oral presentations to one another about their work. In IMP, your assessment of students becomes integrated with learning, and you evaluate students according to a variety of criteria, including class participation, daily homework assignments, Problems of the Week, portfolios, and unit assessments. The IMP *Teaching Handbook* provides many practical suggestions on how to get the best possible results using this curriculum in *your* classroom.

What is in Year 2 of the IMP curriculum?

Year 2 of the IMP curriculum contains five units.

Solve It!

The focus of this unit is on using equations to represent real-life situations and on developing the skills to solve these equations. Students begin with situations used in the first year of the curriculum and develop algebraic representations of problems. In order to find solutions to the equations that occur, students explore the concepts of equivalent expressions and equivalent equations. Using these concepts, they develop principles such as the distributive property for working with algebraic expressions and equations, and they learn methods that they can use to solve any linear equation. They also explore the relationship between an algebraic expression, a function, an equation, and a graph, and they examine how to use graphs to solve nonlinear equations.

Is There Really a Difference?

In this unit, students collect data and compare different population groups to one another. In particular, they concentrate on this question:

If a sample from one population differs in some respect from a sample from a different population, how reliably can you infer that the overall populations differ in that respect?

They begin by making double bar graphs of some classroom data and explore the process of making and testing hypotheses. Students realize that there is variation even among different samples from the same population, and they see the usefulness of the concept of a *null hypothesis* as they examine this variation. They build on their understanding of standard deviation from the Year 1 unit *The Pit and the Pendulum* and learn that the chi-square (χ^2) statistic can give them the probability of seeing differences of a certain size in samples when the populations are really the same. Their work in this unit culminates in a two-week project in which they propose a hypothesis about two populations that they think really differ in some respect. They then collect sample data about the two populations and analyze their data by using bar graphs, tables, and the χ^2 statistic.

Do Bees Build It Best?

In this unit students work on this problem:

> *Bees store their honey in honeycombs that consist of cells they make out of wax. What is the best design for a honeycomb?*

To analyze this problem, students begin by learning about area and the Pythagorean theorem. Then, using the Pythagorean theorem and trigonometry, students find a formula for the area of a regular polygon with fixed perimeter and find that the larger the number of sides, the larger the area of the polygon. Students then turn their attention to volume and surface area, focusing on prisms that have a regular polygon as the base. They find that for such prisms—if they also want the honeycomb cells to fit together—the mathematical winner, in terms of maximizing volume for a given surface area, is a regular hexagonal prism, which is essentially the choice of the bees.

Cookies

The focus of this unit is on graphing systems of linear inequalities and solving systems of linear equations. Although the central problem is one in linear programming, the major goal of the unit is for students to learn how to manipulate equations and how to reason using graphs.

Students begin by considering a classic type of linear programming problem in which they are asked to maximize the profits of a bakery that makes plain and iced cookies. They are constrained by the amount of ingredients they have on hand and the amount of oven and labor time available. First students work toward a graphical solution of the problem. They see how the linear function can be maximized or minimized by studying the graph. Because the maximum or minimum point they are looking for is often the intersection of two lines, they are motivated to investigate a method for solving two equations in two unknowns. They then return to work in groups on the cookie problem, each group presenting both a solution and a proof that their solution does maximize profits. Finally, each group invents its own linear programming problem and makes a presentation of the problem and its solution to the class.

All About Alice

This unit starts with a model based on Lewis Carroll's *Alice in Wonderland,* a story in which Alice's height is doubled or halved by eating or drinking certain foods she finds. Out of the discussion of this situation come the basic principles for working with exponents—positive, negative, zero, and even fractional—and an introduction to logarithms. Building on the work with exponents, the unit discusses scientific notation and the manipulation of numbers written in scientific notation.

How do the four years of the IMP curriculum fit together?

The four years of the IMP curriculum form an integrated sequence through which students can learn the mathematics they will need both for further education and on the job. Although the organization of the IMP curriculum is very different from the traditional Algebra I-Geometry-Algebra II/Trigonometry-Precalculus sequence, the important mathematical ideas are all there.

Here are some examples of how both traditional concepts and topics new to the high school curriculum are developed:

Linear equations

In Year 1 of the IMP curriculum, students develop an intuitive foundation of algebraic thinking, including the use of variables, which they build on throughout the program. In the Year 2 unit *Solve It!,* students use the concept of equivalent equations to see how to solve any linear equation in a single variable. Later in Year 2, in a unit called *Cookies* (about maximizing profits for a bakery), they solve pairs of linear equations in two variables, using both algebraic and geometric methods. In *Meadows or Malls?* (Year 3), they extend those ideas to systems with more than two variables, and see how to use matrices and the technology of graphing calculators to solve such systems.

Measurement and the Pythagorean theorem

Measurement, including area and volume, is one of the fundamental topics in geometry. The Pythagorean theorem is one of the most important geometric principles ever discovered. In the Year 2 unit *Do Bees Build It Best?,* students combine these ideas with their knowledge of similarity (from the Year 1 unit *Shadows*) to see why the hexagonal prism of the bees' honeycomb design is the most efficient regular prism possible. Students also use the Pythagorean theorem in later units, applying it to develop such principles as the distance formula in coordinate geometry.

Trigonometric functions

In traditional programs, the trigonometric functions are introduced in the eleventh or twelfth grade. In the IMP curriculum, students begin working with trigonometry in Year 1 (in *Shadows*), using right-triangle trigonometry in several units in Years 2 and 3 (including *Do Bees Build It Best?*). In the Year 4 unit *High Dive,* they extend trigonometry from right triangles to circular functions, in the context of a circus act in which a performer falls from a Ferris wheel into a moving tub of water. (In *High Dive,* students also learn principles of physics, developing laws for falling objects and using vectors to find vertical and horizontal components of velocity.)

Standard deviation and the binomial distribution

Standard deviation and the binomial distribution are major tools in the study of probability and statistics. *The Game of Pig* gets students started by building a firm understanding of concepts of probability and the phenomenon of experimental variation. Later in Year 1 (in *The Pit and the Pendulum*), they use standard deviation to see that the period of a pendulum is determined primarily by its length. In Year 2, they compare standard deviation with the chi-square test in examining whether the difference between two sets of data is statistically significant. In *Pennant Fever* (Year 3), students use the binomial distribution to evaluate a team's chances of winning the baseball championship, and in *The Pollster's Dilemma* (Year 4), students tie many of these ideas together in the central limit theorem, seeing how the margin of error and the level of certainty for an election poll depend on its size.

Does the program work?

The IMP curriculum has been thoroughly field-tested and enthusiastically received by hundreds of classroom teachers around the country. Their enthusiasm is based on the success they have seen in their own classrooms with their own students. For instance, IMP teacher Dennis Cavaillé says, "For the first time in my teaching career, I see lots of students excited about solving math problems inside *and* outside of class."

These informal observations are backed up by more formal evaluations. Dr. Norman Webb of the Wisconsin Center for Education Research has done several studies comparing the performance of students using the IMP curriculum with the performance of students in traditional programs. For instance, he has found that IMP students do as well as students in traditional mathematics classes on standardized tests such as the SAT. This is especially significant because IMP students spend about twenty-five percent of their time studying topics, such as statistics, not covered on these tests. To measure IMP students' achievement in these other areas, Dr. Webb conducted three separate studies involving students at different grade levels and in different locations. The three tests used in these studies involved statistics, quantitative reasoning, and general problem solving. In all three cases, the IMP students outperformed their counterparts in traditional programs by a statistically significant margin, even though the two groups began with equivalent scores on eighth grade standardized tests.

But one of our proudest achievements is that IMP students are excited about mathematics, as shown by Dr. Webb's finding that they take more mathematics courses in high school than their counterparts in traditional programs. We think this is because they see that mathematics can be relevant to their own lives. If so, then the program works.

Dan Fendel

Diane Resek

Lynne Alper

Sherry Fraser

Note to Students

This textbook represents the second year of a four-year program of mathematics learning and investigation. As in the first year, the program is organized around interesting, complex problems, and the concepts you learn grow out of what you'll need to solve those problems.

These pages in the student book welcome students to the program.

- *If you studied IMP Year 1*

If you studied IMP Year 1, then you know the excitement of problem-based mathematical study, such as devising strategies for a complex dice game, learning the history of the Overland Trail, and experimenting with pendulums.

The Year 2 program extends and expands the challenges that you worked with in Year 1. For instance:

- In Year 1, you began developing a foundation for working with variables. In Year 2, you will build on this foundation in units that demonstrate the power of algebra to solve problems, including some that look back at situations from Year 1 units.
- In Year 1, you used principles of statistics to help predict the period of a 30-foot pendulum. In Year 2, you will learn another statistical method, one that will help you to understand statistical comparisons of populations. One important part of your work will be to prepare, conduct, and analyze your own survey.

You'll also use ideas from geometry to understand why the design of bees' honeycombs is so efficient, and you'll use graphs to help a bakery decide how many plain and iced cookies they should make to maximize their profits. Year 2 closes with a literary adventure—you'll use Lewis Carroll's classic *Alice's Adventures in Wonderland* to explore and extend the meaning of exponents.

- *If you didn't study IMP Year 1*

If this is your first experience with the Interactive Mathematics Program (IMP), you can rely on your classmates and your teacher to fill in what you've missed. Meanwhile, here are some things you should know about the program, how it was developed, and how it is organized.

The Interactive Mathematics Program is the product of a collaboration of teachers, teacher-educators, and mathematicians who have been working together since 1989 to reform the way high school mathematics is taught. About one hundred thousand students and five hundred teachers used these materials before they were published. Their experiences, reactions, and ideas have been incorporated into this final version.

Our goal is to give you the mathematics you need in order to succeed in this changing world. We want to present mathematics to you in a manner that reflects how mathematics is used and that reflects the different ways people work and learn together. Through this perspective on mathematics, you will be prepared both for continued study of mathematics in college and for the world of work.

This book contains the various assignments that will be your work during Year 2 of the program. As you will see, these problems require ideas from many branches of mathematics, including algebra, geometry, probability, graphing, statistics, and trigonometry. Rather than present each of these areas separately, we have integrated them and presented them in meaningful contexts, so you will see how they relate to each other and to our world.

Each unit in this four-year program has a central problem or theme, and focuses on several major mathematical ideas. Within each unit, the material is organized for teaching purposes into "days," with a homework assignment for each day. (Your class may not follow this schedule exactly, especially if it doesn't meet every day.)

At the end of the main material for each unit, you will find a set of supplementary problems. These problems provide you with additional opportunities to work with ideas from the unit, either to strengthen your understanding of the core material or to explore new ideas related to the unit.

Although the IMP program is not organized into courses called "Algebra," "Geometry," and so on, you will be learning all the essential mathematical concepts that are part of those traditional courses. You will also be learning concepts from branches of mathematics—especially statistics and probability—that are not part of a traditional high school program.

To accomplish your goals, you will have to be an active learner, because the book does not teach directly. Your role as a mathematics student will be to experiment, to investigate, to ask questions, to make and test conjectures, and to reflect, and then to communicate your ideas and conclusions both orally and in writing. You will do some of your work in collaboration with fellow students, just as users of mathematics in the real world often work in teams. At other times, you will be working on your own.

We hope you will enjoy the challenge of this new way of learning mathematics and will see mathematics in a new light.

Dan Fendel Diane Resek Lynne Alper Sherry Fraser

Finding What You Need

We designed this guide to help you find what you need amid all the information it provides. Each of the following components has a special treatment in the layout of the guide.

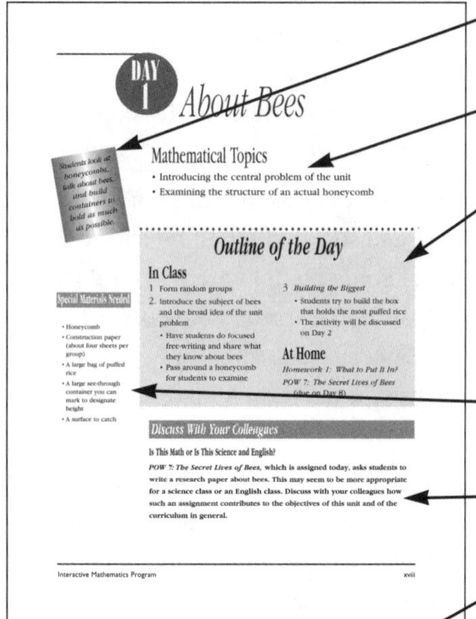
DAY 1 About Bees

Students look at honeycombs, talk about bees, and build containers to hold as much as possible.

Mathematical Topics
- Introducing the central problem of the unit
- Examining the structure of an actual honeycomb

Outline of the Day

In Class
1. Form random groups
2. Introduce the subject of bees and the broad idea of the unit problem
   - Have students do focused free-writing and share what they know about bees
   - Pass around a honeycomb for students to examine
3. Building the Biggest
   - Students try to build the box that holds the most puffed rice
   - The activity will be discussed on Day 2

At Home

Homework 1: What to Put It In?

POW 7: The Secret Lives of Bees (due on Day 8)

Special Materials Needed
- Honeycomb
- Construction paper (about four sheets per group)
- A large bag of puffed rice
- A large see-through container you can mark to designate height
- A surface to catch

Discuss With Your Colleagues

Is This Math or Is This Science and English?

POW 7: The Secret Lives of Bees, which is assigned today, asks students to write a research paper about bees. This may seem to be more appropriate for a science class or an English class. Discuss with your colleagues how such an assignment contributes to the objectives of this unit and of the curriculum in general.

Interactive Mathematics Program xvii

Synopsis of the Day: The key idea or activity for each day is summarized in a brief sentence or two.

Mathematical Topics: Mathematical issues for the day are presented in a bulleted list.

Outline of the Day: Under the *In Class* heading, the outline summarizes the activities for the day, which are keyed to numbered headings in the discussion. Daily homework assignments and Problems of the Week are listed under the *At Home* heading.

Special Materials Needed: Special items needed in the classroom for each day are bulleted here.

Discuss With Your Colleagues: This section highlights topics that you may want to discuss with your peers.

Post This: The *Post This* icon indicates items that you may want to display in the classroom.

Asides: These are ideas outside the main thrust of a discussion. They include background information, refinements, or subtle points that may only be of interest to some students, ways to help fill in gaps in understanding the main ideas, and suggestions about when to bring in a particular concept.

Suggested Questions: These are specific questions that you might ask during an activity or discussion to promote student insight or to determine whether students understand an idea. The appropriateness of these questions generally depends on what students have already developed or presented on their own.

Icons for Student Written Products

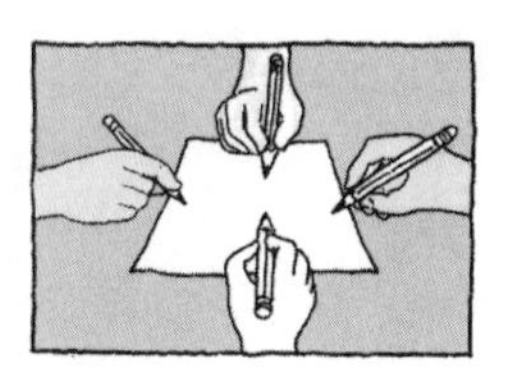
Single Group report

Individual reports

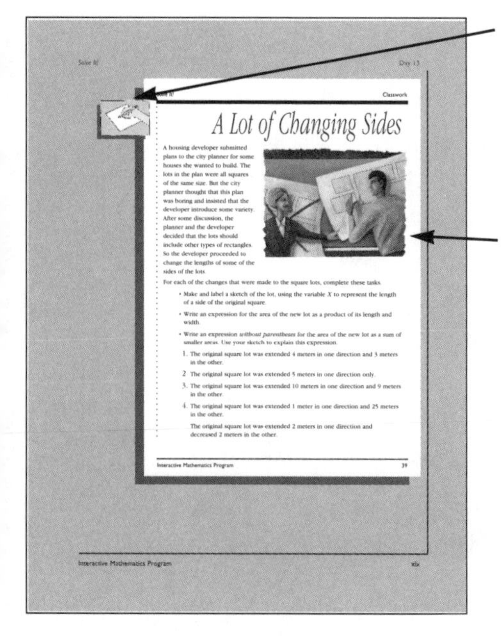
Classwork

A Lot of Changing Sides

A housing developer submitted plans to the city planner for some houses she wanted to build. The lots in the plan were all squares of the same size. But the city planner thought that this plan was boring and insisted that the developer introduce some variety. After some discussion, the planner and the developer decided that the lots should include other types of rectangles. So the developer proceeded to change the lengths of some of the sides of the lots.

For each of the changes that were made to the square lots, complete these tasks.

- Make and label a sketch of the lot, using the variable X to represent the length of a side of the original square.
- Write an expression for the area of the new lot as a product of its length and width.
- Write an expression *without parentheses* for the area of the new lot as a sum of smaller areas. Use your sketch to explain this expression.

1. The original square lot was extended 4 meters in one direction and 3 meters in the other.
2. The original square lot was extended 5 meters in one direction only.
3. The original square lot was extended 10 meters in one direction and 9 meters in the other.
4. The original square lot was extended 1 meter in one direction and 25 meters in the other.

The original square lot was extended 2 meters in one direction and decreased 2 meters in the other.

Interactive Mathematics Program 39

Icons for Student Written Products: For each group activity, there is an icon suggesting a single group report, individual reports, or no report at all. If graphs are included, the icon indicates this as well. (The graph icons do not appear in every unit.)

Embedded Student Pages: The teacher guide contains reduced-size copies of the pages from the student book, including the "transition pages" that appear occasionally within each unit to summarize each portion of the unit and to prepare students for what is coming. The reduced-size classwork and homework assignments follow the teacher notes for the day on which the activity is begun. Having all of these student pages in the teacher's guide is a helpful way for you to see things from the students' perspective.

Additional Information

Here is a brief outline of other tools we have included to assist you and make both the teaching and the learning experience more rewarding.

Glossary: This section, which is found at the back of the book, gives the definitions of important terms for all of Year 2 for easy reference. The same glossary appears in the student book.

Appendix A: Supplemental Problems: This appendix contains a variety of interesting additional activities for the unit, for teachers who would like to supplement material found in the regular classroom problems. These additional activities are of two types—*reinforcements,* which help increase student understanding of concepts that are central to the unit, and *extensions,* which allow students to explore ideas beyond the basic unit.

Appendix B: Blackline Masters: For each unit, this appendix contains materials you can reproduce that are not available in the student book and that will be helpful to teacher and student alike. They include the end-of-unit assessments as well as such items as diagrams from which you can make transparencies. Semester assessments for Year 2 are included in *Do Bees Build It Best?* (for first semester) and *All About Alice* (for second semester).

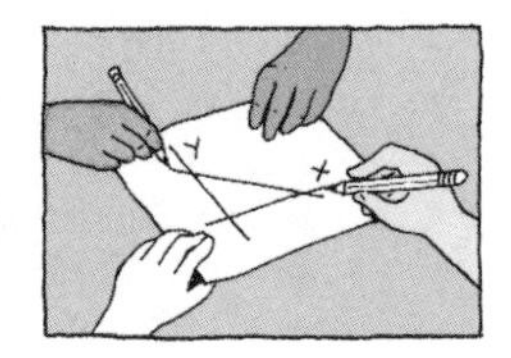
Single group graph

Individual graphs

No report at all

Year 2 IMP Units

Solve It!

Is There Really a Difference?

Do Bees Build It Best?

Cookies **(in this book)**

All About Alice

Contents

"Cookies" Overview

Summary of the Unit

The focus of this unit is using graphs of linear equations and inequalities to understand and solve problems. Although the central unit problem is a linear programming problem, the goal of the unit is not to have students learn an algorithm for solving such problems (for instance, "the solution is always at a corner"). Rather, the goals are for students to deepen their understanding of the relationship between equations or inequalities and their graphs and to reason and solve problems using graphs.

Students begin by considering a typical linear programming problem—how to maximize the profit of a bakery that makes plain and iced cookies. The situation is constrained by the amount of cookie dough and icing the bakery has on hand and the amount of oven time and labor time available. Each of these constraints represents a linear inequality affecting the number of cookies of each type to be made. The profit is a linear function of these numbers of cookies.

Students work toward a graphical solution of the problem. They begin with an intuitive investigation of linear inequalities and their graphs, leading to the recognition that the graph of a linear inequality in two variables is a half plane, bounded by the graph of the related linear equation.

Next, students combine the linear inequalities that represent constraints in the unit problem. They see that the ordered pairs representing the numbers of plain and iced cookies must be inside or on the boundary of a polygonal region in the coordinate plane. This region is called the **feasible region** for the set of constraints. Students realize that their goal is to find the point or points in this region where the linear profit function has its maximum value.

They then leave the cookie problem for a while and work on problems with fewer constraints. They study the linear profit function (or its analog) by looking at the points where this profit function (in two variables) has a particular fixed value. They discover that for any choice of profit, these points lie on a straight line, often called a **profit line.** They see that as they change the profit, they also change the line, but that the line "keeps its direction"; that is, all such profit lines are parallel to each other. (Students are introduced to the term *slope* as referring to the direction of a line, without a formal definition. A numerical value for the slope of a line is not needed in this unit; the formal definition of slope is developed by students in the Year 3 unit *Small World, Isn't It?*)

From this idea of a family of parallel profit lines, students conclude that to maximize the profit, they need the line in the family that intersects their feasible region at some extreme. Students see that in most cases, such an extreme must be the point of intersection of two lines defining the boundary of the feasible region. This leads students to investigate methods for solving two equations in two unknowns. In an activity extending over three days, groups develop their own methods for solving these systems of equations.

These general ideas about linear programming problems and families of parallel lines are explored through a variety of problem settings, including an artist deciding what type of pictures to paint, a pet owner choosing food for his pet, and a university deciding how many in-state and how many out-of-state students to admit.

Students then return to solve the original cookie problem, using both their experience with other linear programming problems and the tools they have developed for solving systems of equations. They are asked to present a clear argument explaining how they know that their solution does maximize profits.

The final activity of the unit is a project in which each group creates its own linear programming problem and presents the problem and its solution to the class. This activity helps students to solidify their understanding of linear programming problems. By manipulating the numbers in the constraints so that the solution to their problem makes sense, students gain a deeper understanding of linear equations and inequalities.

Note: Many ideas in this unit will be extended in the Year 3 unit *Meadows or Malls?* in which the central problem is a linear programming situation in six variables. Students begin that unit by extending to three dimensions their understanding of how to find graphical solutions to two-dimensional linear programming problems. They then generalize what they have learned about graphical solutions to a more algebraic form, which allows them to apply the ideas to problems in more than three dimensions for which they cannot use graphs. Finally, they develop concepts about matrices and matrix operations, and use matrices to solve the complex systems of equations involved in that unit's central problem.

This outline summarizes the unit's overall organization and some of the main activities.

- Days 1–3: Introduction of the unit problem and representation of constraints by inequalities; manipulating inequalities algebraically
- Days 4–5: Graphing linear inequalities in two variables and seeing that the graph of a linear inequality is a half plane

- Days 6–7: Combining graphs of linear inequalities, and introduction of the concept of a feasible region
- Days 8–13: Use of a family of parallel profit lines to find the maximum (or minimum) value of a linear function on a feasible region
- Day 14: Using a graphing calculator to understand feasible regions and the "family of parallel lines" reasoning
- Days 15–18: Developing methods for solving systems of equations, and introduction to the concepts of inconsistent and dependent equations
- Days 19–22: Solving the unit problem and another complex linear programming problem
- Days 23–27: Creating and presenting linear programming problems
- Days 28–29: Portfolios, end-of-unit assessments, and summing up

Concepts and Skills

Here are the concepts and skills that students will encounter during the course of this unit, summarized by category.

Using variables to represent problems

- Expressing and interpreting constraints using inequalities
- Expressing problem situations using systems of linear equations

Working with variables, equations, and inequalities

- Finding equivalent equations or inequalities
- Solving linear equations for one variable in terms of another
- Developing and using a method for solving systems of two linear equations in two unknowns
- Recognizing inconsistent and dependent equations

Graphing

- Graphing individual linear inequalities and systems of linear inequalities
- Finding the equation of a straight line and the inequality for a half plane
- Using graphing calculators to draw feasible regions

- Relating the idea of intersection of graphs to the idea of common solution of equations
- Using graphing calculators to estimate coordinates of points of intersection of graphs

Reasoning based on graphs

- Seeing that setting a linear expression equal to different constants gives a family of parallel lines
- Finding the maximum or minimum of a linear function over a region
- Examining how the parameters in a problem affect the solution
- Combining various concepts and skills listed above to solve linear programming problems with two variables

Creating word problems

- Creating problems that can be solved using two equations in two unknowns
- Creating a problem that can be solved by linear programming methods

Other topics will arise in connection with the Problems of the Week.

Materials

You will need to provide these materials during the course of the unit (in addition to standard equipment and materials such as graphing calculators, transparencies, chart paper, and marking pens).

- Graph paper transparencies
- Grid chart paper
- Colored pencils (extras in case students forget them)
- Pieces of colored paper (for introducing *POW 11: A Hat of a Different Color*)
- A transparency of the feasible region for *Homework 7: Picturing Pictures* (see Appendix B)
- (Optional) Copies of *Grading Programming Problem Presentations* for use on Days 26 and 27 (see Appendix B)

Students will need to provide these materials:

- Scientific calculator (for use at home)
- Graph paper (for use at home)
- Colored pencils (four different colors)
- Rulers

Grading

The IMP *Teaching Handbook* contains general guidelines about how to grade students in an IMP class. You will probably want to check daily that students have done their homework and include the regular completion of homework as part of students' grades. Your grading scheme will probably also include Problems of the Week, the unit portfolio, and the end-of-unit assessments.

Because you will not be able to read thoroughly every assignment that students turn in, you will need to select certain assignments to read carefully and to base grades on. Here are some suggestions.

- Part I of *Homework 4: Inequality Stories*
- *Profitable Pictures* (handed in on Day 9)
- *Homework 11: Changing What You Eat*
- *Get the Point* write-ups and presentations (Day 17)
- *Homework 17: A Reflection on Money*
- *"How Many of Each Kind?" Revisited* write-ups and presentations (Day 21)
- *Producing Programming Problems* write-ups and presentations (Days 26 and 27)

If you want to base your grading on more tasks, there are many other homework assignments, class activities, and oral presentations you can use.

Interactive Mathematics Program

Integrated High School Mathematics

Cookies

Cookies

Cookies and Inequalities

This page in the student book introduces Days 1 through 3.

Emily Perez and Kim Bell discuss homework results within their group.

The central problem of this unit involves helping a bakery to maximize its profits. The problem is complex. In the opening days of the unit, your main task is to organize all of the information and express the bakery's situation in algebraic terms, using inequalities and linear expressions.

Interactive Mathematics Program 2

Students try to satisfy several constraints and to maximize profits.

Mathematical Topics

- Introducing the central unit problem
- Finding numbers that satisfy constraints

Outline of the Day

In Class

1. Form random groups
2. *How Many of Each Kind?*
    - Introduce the unit problem
    - Students look for a combination that fits all the constraints
    - The activity will be discussed on Day 2

At Home

Homework 1: A Simpler Cookie

1. Forming Groups

At the beginning of the unit, place students in new groups as described in the IMP *Teaching Handbook.* We recommend that you create new groups again on Day 15.

2. *How Many of Each Kind?*

Have students read *How Many of Each Kind?* (You may wish to have several students take turns reading portions of it aloud.) Tell them that this is the main problem of the unit (we will sometimes refer to it simply as "the unit problem" or "the cookie problem"), but that today's work will be just a preliminary look at it. This activity will be discussed tomorrow.

Note: Because students will return to this problem at the end of the unit, you may want to save the results that they get now.

- *Getting started*

Have students work in groups on the questions in the activity. As they work, give several groups overhead transparencies and pens to use in preparing presentations of their results for tomorrow's discussion. Look for groups that organized the information in different ways.

Part of the difficulty of this problem is keeping track of all the numbers. Let students develop their own way of organizing the information, because this is a valuable skill, both in the classroom and in life. Although this approach may require more class time, it is time well spent.

As you observe the groups at work, make sure students realize that the numbers they are looking for are in *dozens* of each kind. For example, they should give answers as "4 dozen plain, 3 dozen iced," not "48 plain, 36 iced."

Students may have trouble at first because they think they have to make use of all the ingredients, all the oven space, and all the preparation time available. Let them become aware on their own that this is not only impossible but also not required by the problem.

Possible hint: "Could the Woos make 1 dozen of each kind? What about 3 dozen plain and 5 dozen iced?" and so on.

If groups seem to be stuck, ask them if it's possible for the Woos to make 1 dozen of each kind, or 3 dozen plain and 5 dozen iced, and so on. In other words, urge them to use a guess-and-check approach. Do not push them to think about the problem in any different or "better" way.

Homework 1: A Simpler Cookie

This homework represents a simpler version of the Woos' cookie problem. You might remind students that in mathematics such simplifications often help show how to attack a particular type of problem. Also, tell students to bring colored pencils (at least four distinct colors) for use beginning tomorrow. They will need them occasionally throughout the unit.

How Many of Each Kind?

Abby and Bing Woo own a small bakery that specializes in cookies. They make only two kinds of cookies—plain and iced. They need to decide *how many dozens* of each kind of cookie to make for tomorrow.

The Woos know that each dozen of their *plain* cookies requires 1 pound of cookie dough (and no icing), and each dozen of their *iced* cookies requires 0.7 pounds of cookie dough and 0.4 pounds of icing. The Woos also know that each dozen of the plain cookies requires about 0.1 hours of preparation time, and each dozen of the iced cookies requires about 0.15 hours of preparation time. Finally, they know that no matter how many of each kind they make, they will be able to sell them all.

The Woos' decision is limited by three factors.

- The ingredients they have on hand—they have 110 pounds of cookie dough and 32 pounds of icing.
- The amount of oven space available—they have room to bake a total of 140 dozen cookies for tomorrow.
- The amount of preparation time available—together they have 15 hours for cookie preparation.

Continued on next page

Why on earth should the Woos care how many cookies of each kind they make? Well, you guessed it! They want to make as much profit as possible. The plain cookies sell for $6.00 a dozen and cost $4.50 a dozen to make. The iced cookies sell for $7.00 a dozen and cost $5.00 a dozen to make.

The Big Question is:

> *How many dozens of each kind of cookie should Abby and Bing make so that their profit is as high as possible?*

1. a. To begin answering the Big Question, find one combination of dozens of plain cookies and dozens of iced cookies that will satisfy all of the conditions in the problem.

   b. Next, find out how much profit the Woos will make on that combination of cookies.

2. Now find a different combination of dozens of cookies that fits the conditions but that yields a greater profit for the Woos.

This problem was adapted from one in *Introduction to Linear Programming, 2nd Edition,* by R. Stansbury Stockton, Allyn and Bacon, 1963, pp. 19–35.

Homework 1 A Simpler Cookie

The Woos have a rather complicated problem to solve. Let's make it simpler. Finding a solution to a simpler problem may lead to a method for solving the original problem.

Assume that the Woos still make both plain and iced cookies, and that they still have 15 hours altogether for cookie preparation. But now assume that they have an unlimited amount of both cookie dough and icing, and that they have an unlimited amount of space in their oven.

The other information is unchanged.

- Preparing a dozen plain cookies requires 0.1 hours.
- Preparing a dozen iced cookies requires 0.15 hours.
- The plain cookies sell for $6.00 a dozen.
- It costs $4.50 a dozen to make plain cookies.
- The iced cookies sell for $7.00 a dozen.
- It costs $5.00 a dozen to make iced cookies.

As before, the Woos know that no matter how many of each kind they make, they will be able to sell them all.

1. Find at least five combinations of plain and iced cookies that the Woos could make without working more than 15 hours. For each combination, find their profit.
2. Find the combination of plain and iced cookies that you think would give the Woos the greatest profit. Explain why you think no other combination will yield a greater profit.

DAY 2 The Constraints

Students express the constraints symbolically as inequalities.

Mathematical Topics

- Finding numbers that satisfy constraints
- Expressing constraints stated in words as algebraic inequalities
- Expressing profit as an algebraic expression

Outline of the Day

In Class

1. Discuss *Homework 1: A Simpler Cookie*
2. Discuss *How Many of Each Kind?* (from Day 1)
   - Groups present different ways to organize the data
   - Introduce the term **constraint**
3. Constraints as inequalities
   - Introduce the constraints first verbally and then symbolically as inequalities
   - Post the constraint inequalities
   - Discuss the profit expression and distinguish it from the constraints
4. Discuss restrictions on P and I if students raise the issue

At Home

Homework 2: Investigating Inequalities

1. Discussion of *Homework 1: A Simpler Cookie*

Have students discuss their homework solutions briefly in groups. Then ask the spade card members from one or two groups to present the group's solution. The fact that there is only one constraint (preparation time) makes this problem much simpler than the original.

Students do not need to get a definitive resolution to this problem. That is, they don't need to be able to prove, or even be sure, that a particular choice is optimal. The purpose of the assignment is for students to get their feet wet. They will be learning how to deal with problems of this type over the course of the unit.

Probably at least some students will see that in order to maximize profit, one should maximize the number of plain cookies. In other words, the Woos should make 150 dozen plain cookies and no iced cookies. Students may arrive at this conclusion in various ways, although they may be somewhat vague about why this solution is best. Some may simply use guess-and-check, seeing that the more plain cookies the Woos make, the higher the profit seems to be.

Note: In 15 hours, the Woos can make 150 dozen plain cookies, which will yield a profit of $225. By comparison, in 15 hours, they can make 100 dozen iced cookies, which will yield a profit of only $200.

2. Discussion of *How Many of Each Kind?*

You can begin the discussion with presentations by the groups to which you gave overhead transparencies yesterday. You might have the heart card members of these groups present their organizational schemes for keeping track of the various conditions and deciding what profit would result from different cookie combinations. Then have heart card members of other groups offer other possible combinations of cookies. You can use one of the schemes presented (or a blend that the class develops from them) to keep track of students' suggestions.

"Are you sure that this combination fits all the conditions? How do you know?"

As combinations are suggested, ask students to check whether the combinations satisfy the conditions by calculating the amount of cookie dough, icing, oven space, and preparation time required for that combination, and determining if the results are within the conditions of the problem. (*Note:* The amount of oven space used is determined indirectly by the number of dozens of cookies made.)

Introduce the term **constraint** in this context as a synonym for *condition* (especially in the sense of *restrictive* condition). You might remind students that they saw this term when they were forming families in the Year 1 unit *The Overland Trail.*

Ask students how and when to compute the profit for each combination. They may recognize that it makes sense to wait until they can establish whether a combination fits all the constraints before they make the calculation. Leave them in charge of the process as much as possible.

"What's an example of a combination that does not fit all of the constraints?"

Ask groups for a combination they tried that *did not* satisfy the constraints, and have them explain which constraint or constraints the combination failed to satisfy. It is just as important for students to see why a combination is excluded as it is for them to show that it is included. It is also important for students to see that each condition is a separate constraint and that a combination must satisfy all four constraints.

Do enough of these combinations to develop a clear method for working with them. You do not need to look at all the combinations that students suggest.

3. Constraints as Inequalities

"How did you decide whether a combination fit this constraint?"

Pick one of the conditions to focus on, such as the amount of cookie dough available, and ask students to explain how they decided whether a combination they tried satisfied this constraint. As a class, develop a statement that tells when a combination does fit that constraint. The statement might go something like this:

> "You take the number of dozens of iced cookies, multiply that by 0.7, and add the number of dozens of plain cookies, and you can't get more than 110."

You may have to start with a less detailed statement than this one and gradually get students to refine it. Questions such as, "Do we want to multiply by the number of cookies?" can be used to prompt refinements such as talking about the number of *dozens* of cookies.

"How can you express this constraint symbolically?"

Then ask students how they might express this "cookie dough constraint" symbolically. If a further hint is needed, suggest that they choose variables to represent "number of dozens of plain cookies" and "number of dozens of iced cookies." We will use P and I, respectively, for these two variables. Using these variables, students should see that the cookie dough constraint can be expressed by the inequality $P + 0.7I \leq 110$.

Then have students work in groups to write verbal as well as symbolic statements for the other constraints. It is important that all students be able to deal with both ways of expressing the condition. You may want to randomly ask some groups to present verbal expressions and others to present symbolic expressions. The class should end up with a set of constraint inequalities that looks something like this:

$P + 0.7I \leq 110$ (for the amount of cookie dough)

$0.4I \leq 32$ (for the amount of icing)

$P + I \leq 140$ (for the amount of oven space)

$0.1P + 0.15I \leq 15$ (for the amount of the Woos' preparation time)

Have students record these constraint inequalities on chart paper. Post this constraint chart for later use, perhaps labeling it "Cookies Constraints."

Note: The fact that these constraints involve *linear* expressions will be brought out on Day 5 as part of the discussion of *Picturing Cookies—Part I,* although students may raise the issue sooner.

Point out that each of the constraints is represented symbolically by an **inequality.** Use the inequalities to check that the combinations used earlier really do satisfy the constraints. In doing so, students will be repeating their earlier computations. But they should still do it at least once to confirm that the symbolic inequalities are saying the same thing as the verbal expressions of these conditions.

- *The profit expression*

Ask students to develop an algebraic expression for the profit. If they have difficulty with this, have them go back to the chart in which they computed profit for various combinations. They should get the expression $1.5P + 2I$.

Students often make the mistake of treating the expression for the profit as another constraint. You may want to ask why this expression wasn't included in the earlier list of constraints. As needed, bring out the difference between an *expression,* such as $1.5P + 2I$, and a *condition* or *constraint,* such as $P + 0.7I \leq 110$, which is a *statement about* an expression.

4. Restrictions on *P* and *I*

Because of the problem context from which the variables come, it makes sense to insist that P and I be whole numbers. This issue is important, and it definitely should be discussed at this time *if introduced by a student.* Otherwise, we recommend that you ignore it for now. The issue will be more engaging and meaningful for students if it comes initially from them or arises in the natural context of discussions of graphs or feasible regions later in the unit. The unit provides an opportunity for discussing restrictions on the variables in different contexts on Day 6.

If this issue does come up now, you can use these ideas.

- *Two aspects to the restrictions issue*

There are actually two separate issues here:

- that P and I can't be negative
- that P and I must be integers. (Actually, allowing P and I to be multiples of $\frac{1}{12}$ would also make sense.)

It turns out that there is a simple fix for the first of these issues but not for the second.

If students raise either of these issues, we suggest that you broaden the scope of the discussion to include both, perhaps asking students if there are other restrictions on the "eligible" values for P and I. You might point out that the question of eligibility will depend on the particular problem situation. That is, negative or non-integer solutions will make sense in some problems but not in others.

• *Avoiding negative values*

Concerning the restriction to non-negative numbers, you might ask students if they can make this restriction using additional constraints. If a further hint is needed, ask if they can express the condition of "not being negative" as an inequality. They should see that the simple inequalities $P \geq 0$ and $I \geq 0$ will accomplish this. In other words, adding these two constraints to the list fixes this aspect of the model. The class should include these inequalities on its posted constraint chart.

• *Avoiding noninteger values*

The issue of avoiding noninteger values is much more complex, because it cannot be handled by additional inequalities. Perhaps the best approach here is simply to tell students that there is no easy way to handle it and that they should ignore it for now. Emphasize that this means they will need to be especially careful later on to check whether their solution makes sense. If it turns out that the solution that provides the maximum profit is somehow "ineligible," students will have to decide where to go from there.

Note: It turns out in this problem that the combination with the maximum profit does have whole-number values for P and I. The complex issue of what to do if one gets noninteger solutions is discussed in the Year 3 unit *Meadows or Malls?*

You can take this opportunity to talk about mathematical modeling. Bring out that we often need to simplify or ignore certain aspects of a problem in order to get a usable mathematical description. You can introduce the term **mathematical model** for this abstract description of the real-world situation. The Year 4 unit *The World of Functions* will focus on this concept.

Finally, bring out that when simplifications are made, it becomes especially important to refer back to the actual problem after the mathematical analysis is completed.

Homework 2: Investigating Inequalities

In this assignment, students will investigate the question, "What manipulations of an inequality will preserve its truth?" They will be able to use what they learn to simplify the constraints in the central unit problem, as well as other inequalities that arise.

Students also are introduced to number-line graphs for inequalities with one variable. The graphs will be used to justify the conclusions about manipulating inequalities.

• *Reviewing the meaning of inequalities*

In order to do this and other assignments involving inequalities, students need to know how to use the terms and symbols for inequalities for both positive and negative numbers. For example, they need to know that -8 is less than -6 and be able to recognize that the statement $5 \geq -7$ is true.

If your students need a review of the ideas and the notation, you will probably want to use the number line as the basic reference point for the review. (This material was introduced on Day 13 of the Year 1 unit *Patterns*.)

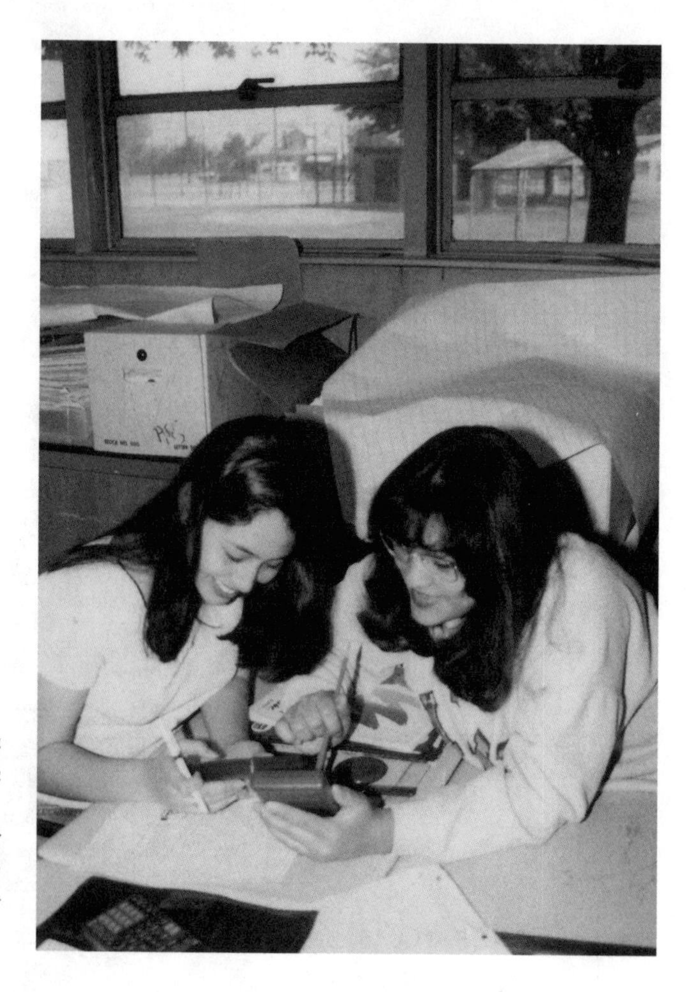

After writing a constraint as an algebraic inequality, Jennifer Rodriguez and Karla Viramontes find numbers that will satisfy it.

Homework 2 Investigating Inequalities

Part I: Manipulating Inequalities

In *Solve It!* you used the mystery bags game to think about ways to change equations but keep them true. For instance, if you had a true equation—that is, two expressions that were equal—you could add the same quantity to both sides of the equation, and the resulting expressions would still be equal.

For example, the statement $3 + 8 = 5 + 6$ is true, because $3 + 8$ and $5 + 6$ are both equal to 11. If you add 7 to both sides, the resulting statement is $3 + 8 + 7 = 5 + 6 + 7$ and this statement is also true.

1. The first aspect of Part I is to investigate whether similar principles hold true for inequalities. Start with the inequality $4 > 3$, which is true. For this inequality, perform each of these tasks and then examine whether the resulting statements are true.

   - Add the same number to both sides of the inequality.
   - Subtract the same number from both sides of the inequality.
   - Multiply both sides of the inequality by the same number.
   - Divide both sides of the inequality by the same number.

   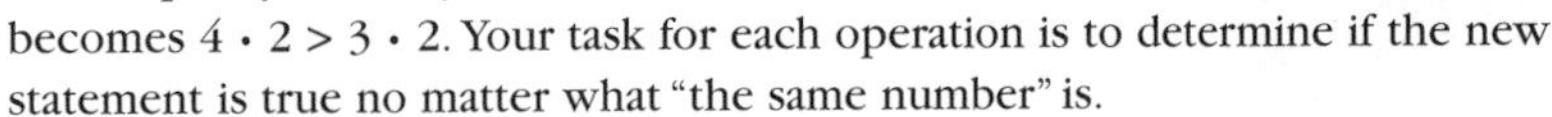

   For example, if you multiply both sides of the inequality $4 > 3$ by 2, the statement becomes $4 \cdot 2 > 3 \cdot 2$. Your task for each operation is to determine if the new statement is true no matter what "the same number" is.

   Try different possibilities for "the same number," using both positive and negative values.

2. After you finish working with the inequality $4 > 3$, start with a different true inequality and see whether you reach the same conclusions.

3. When you are done exploring, state your conclusions. Make them as general as possible.

Continued on next page

Part II: Graphing Inequalities

If an inequality has a single variable in it, we can picture all the numbers that make the inequality true by shading them on a number line. This is called the **graph of the inequality.** An inequality using $<$ or $>$ is called a **strict** inequality. An inequality using $\leq$ or $\geq$ is called **nonstrict.**

For example, the colored portion of this number line represents the graph of the strict inequality $x < 4$:

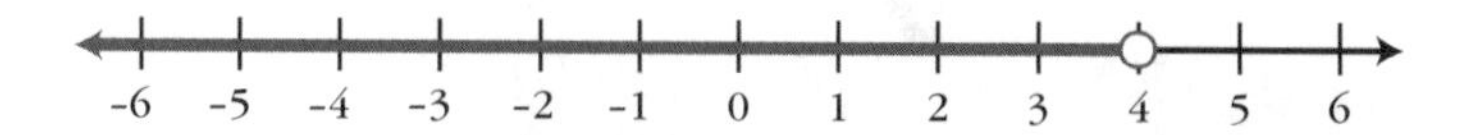

The open circle at the number 4 on the number line means that the number 4 is not included in the graph. (The number 4 is not included because substituting 4 for x gives a false statement.) *Note:* The exclusion of an endpoint is sometimes represented by a parenthesis instead of the open circle.

If we want to include a particular number as part of the graph, we mark that point with a filled-in circle (or by a bracket). For example, the colored portion of the next diagram represents the graph of the nonstrict inequality $x \leq 4$:

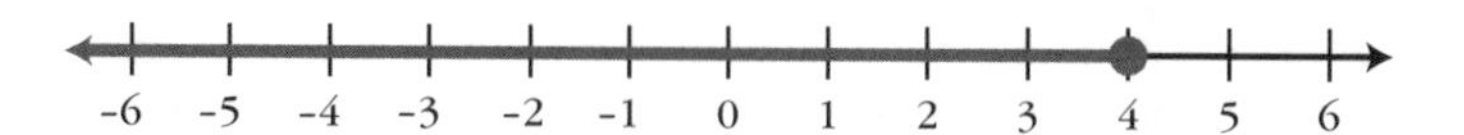

4. Draw the graph of the inequality $x > -2$.
5. Draw the graph of the inequality $x \leq 0$.
6. What inequality goes with this graph?

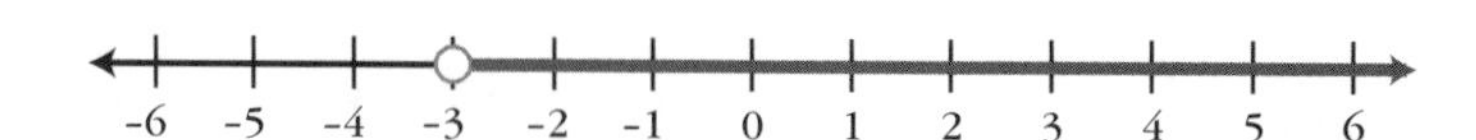

7. How would you use inequalities to describe this graph?

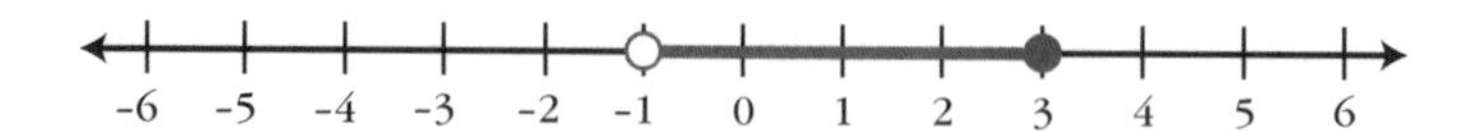

DAY 3

The World of Inequalities

Students work with equivalent inequalities.

Mathematical Topics

- Manipulating inequalities algebraically
- Representing inequalities in one variable by graphs

Outline of the Day

In Class

1. Discuss *Homework 2: Investigating Inequalities*
   - List algebraic manipulations that preserve the truth of inequalities
   - See why multiplying by a negative number reverses the direction of an inequality
2. Variables in inequalities
   - See that rules for manipulating numerical inequalities also apply to inequalities involving variables
3. *My Simplest Inequality*
   - Students find simpler equivalent inequalities
4. Discuss *My Simplest Inequality*
   - Bring out that there is often more than one way to simplify an inequality, and not necessarily a "best" final form

At Home

Homework 3: Simplifying Cookies

1. Discussion of *Homework 2: Investigating Inequalities*

You may want to go over Part II of the homework first, because number lines can be used to explain the rules for preserving the truth of inequalities. Ask volunteers to show their answers to Questions 4 through 7.

Students may have described the graph in Question 7 using two separate inequalities, $x > -1$ and $x \leq 3$. If so, tell them that we usually abbreviate such pairs by combining them into a single inequality, written as $-1 < x \leq 3$.

Note: It's easy to get bogged down in distinctions between "and" and "or," and we recommend that you avoid a discussion of formal logic here. You might suggest that students simply read the inequality $-1 < x \leq 3$ as "x is between -1 and 3, not including -1 and including 3."

• *Part I: Manipulating Inequalities*

Have students discuss in groups the conclusions they reached in Part I. When you think the groups are ready, conduct a whole-class discussion. Have students develop a list of manipulations that can be done to a true inequality to produce inequality statements that are still true.

The operations of addition and subtraction should be easy, but the operations of multiplication and division may be more difficult. For example, students might overlook the case of multiplying or dividing by a negative number, or they might simply say that in this case, the resulting inequality statement is false. You can use ideas in the subsection "Multiplying or dividing by negative numbers" to elicit the rules for multiplication and division and ideas from the subsection "Explaining the reversal" to help students explain these rules.

The students' eventual list should state that if they start with a true inequality statement, they get true statements when they do any of these operations.

- Add the same number to both sides of the inequality.
- Subtract the same number from both sides of the inequality.
- Multiply both sides of the inequality by the same positive number.
- Divide both sides of the inequality by the same positive number.
- Multiply both sides of the inequality by the same negative number and reverse the direction of the inequality.
- Divide both sides of the inequality by the same negative number and reverse the direction of the inequality.

The list might also state explicitly that if they multiply or divide by a negative number and do not reverse the inequality, the result is a false statement.

You should post the class list once students have completed it.

• *Multiplying or dividing by negative numbers*

The difficult cases, of course, are those that involve multiplying or dividing an inequality by a negative number. This is where the rules for working with inequalities differ from those for equations and where the confusion usually sets in.

If needed: "What happens if you multiply both sides of 4 > 3 by -2?"

If students have followed the instruction in the assignment to consider both positive and negative numbers, they will have seen this complication. As an alternative, you can bring it up if students make a general statement like, "You can multiply both sides of the inequality $4 > 3$ by the same thing, and the result is still true." In one way or another, they should be confronted with the question of what happens if they multiply both sides of an inequality by a negative number. (The case of dividing is similar and so won't be discussed separately here.)

Students may simply state that multiplying by a negative number gives a false statement. For example, they may say that "$4(-2) > 3(-2)$" is not correct, because -8 is not more than -6. If they do, ask whether there is any way to adjust the statement "$4(-2) > 3(-2)$" to make it true. As a hint, suggest changing the inequality symbol, or ask, "How *is* -8 related to -6?" Using examples as needed, bring out that they will get a true statement if they reverse the direction of the inequality.

Building on such examples, students should be able to make appropriate general statements about multiplying or dividing inequalities by negative numbers. They should conclude that if they start with a true inequality statement, they get true statements if they do either of these two things:

- Multiply both sides of the inequality by the same negative number and reverse the direction of the inequality.
- Divide both sides of the inequality by the same negative number and reverse the direction of the inequality.

• *Explaining the reversal*

"Why does multiplying by a negative number reverse the inequality?"

Because the reversal of inequalities is a departure from what happens with equations, it's important for students to understand why this happens. Ask students for their own explanations, and follow up as needed.

One approach to explaining the reversal of inequalities is based on the number line. You can begin with an inequality such as $8 > 5$ and ask students to explain this statement in terms of the number line. They should simply point out that 8 is to the right of 5 on the number line.

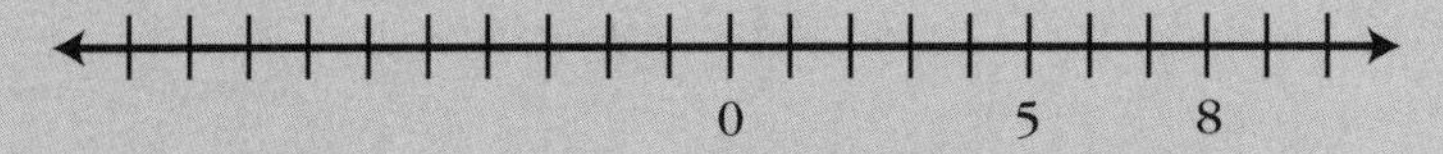

You can ask for a volunteer to show what adding the same thing to both sides of the inequality does. Students should see that it shifts the two numbers, 8 and 5, equal distances (to the right if the addend is positive, and to the left if the addend is negative). For example, they might illustrate adding 2 to both sides by the diagram shown here, and use the diagram to explain that the resulting inequality, $10 > 7$, is also true.

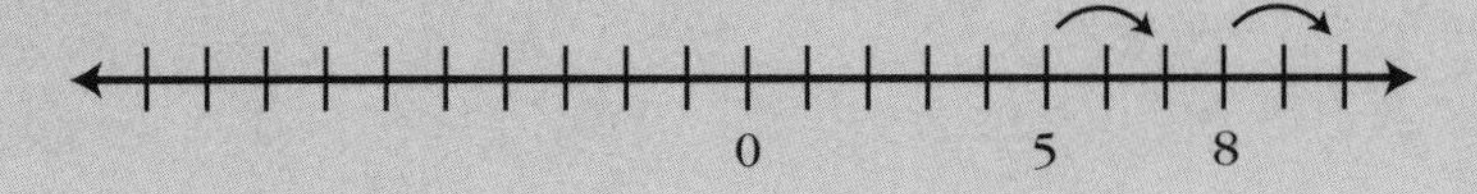

Then ask them to *multiply* both sides of the original inequality, $8 > 5$, by 2. They will see that both numbers move to the right, with $8 \cdot 2$ ending up to the right of $5 \cdot 2$, preserving the relationship.

Next, have students multiply each side of the inequality by -1 and describe what happens in terms of the number line. They might use a phrase like "on the other side of 0" to describe where -5 and -8 are found.

Bring out the symmetry of the situation around 0, and focus on the idea that because 8 is farther to the right than 5, its reflection around 0 ends up farther to the left. The discussion might produce a diagram something like this:

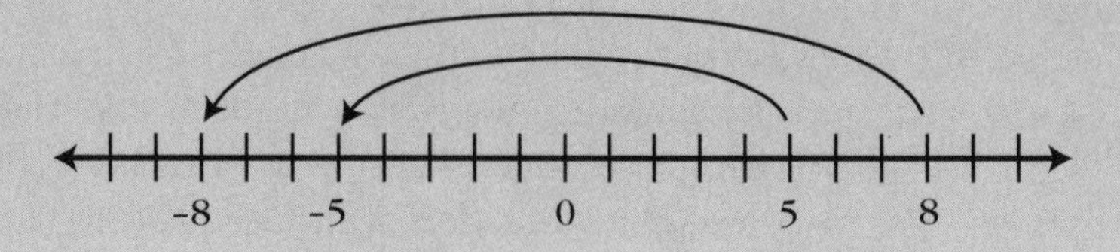

2. From Numbers to Variables

"What do you call two equations with the same solution or solutions?"

Remind students that in *Solve It!* they applied principles similar to those just discussed in order to replace an equation with a simpler one having the same solution (or solutions). Ask what such equations are called. Then review the term **equivalent equations,** and tell students that a similar concept applies for inequalities.

You might begin with a simple inequality such as $x + 3 < 9$. Ask students what they can do to get an inequality with the same solutions as this. They will likely suggest subtracting 3 from both sides to get the inequality $x < 6$. They can test this by finding numbers that fit either $x + 3 < 9$ or $x < 6$ and verifying that such numbers in fact fit the other inequality.

Bring out that the goal for inequalities with one variable is to get a statement that precisely describes the numbers that fit the inequality. For example, the fact that $x < 6$ is equivalent to $x + 3 < 9$ tells us that the solutions to $x + 3 < 9$ are all numbers less than 6 (and only those numbers).

• *Two or more variables*

Also give students an example with two variables, such as $0.5x \geq 2y - 7$. Then ask what simpler inequality they can write that is equivalent to this. For instance, they might multiply both sides by 2 to get $x \geq 4y - 14$ and then add 14 to both sides to get $x + 14 \geq 4y$.

Bring out that there is no single best form for this inequality, because there is no equivalent that "gives the solution" as there is in the case of one variable. For inequalities with two or more variables, the goal of using equivalent inequalities is to find a form that is comparatively easy to work with.

3. *My Simplest Inequality*

With this general introduction, let students work in groups on *My Simplest Inequality*. As they work on this, circulate to see that they are finding "the solution" in Part I and getting a variety of equivalent forms in Part II.

4. Discussion of *My Simplest Inequality*

• *Part I: One Variable Only*

Have students share their work on the examples in Question 1. Bring out that there is often more than one way to simplify a given inequality. For example, here are two different sequences of steps for simplifying the inequality $4 - 2x > 7 + x$ (Question 1d).

$4 - 2x > 7 + x$	$4 - 2x > 7 + x$
$-2x > 3 + x$	$4 > 7 + 3x$
$-3x > 3$	$-3 > 3x$
$x < -1$	$-1 > x$

Some students may prefer to keep the variable on the left side of the inequality (as in the first sequence of steps). Others may prefer to have a positive coefficient for x and avoid the issue of reversing the inequality (as in the second sequence). In either case, be sure students see that the final results, $x < -1$ and $-1 > x$, are equivalent and that neither is necessarily "better" than the other.

• *Part II: Two or More Variables*

On Question 2, you might have one or two students present their examples for Questions 2a and 2c, showing that the examples do or do not fit the original inequality, as required. Be sure students see the distinction between knowing that the two inequalities have solutions in common and actually proving the equivalence. Bring out that it would be impossible for them to verify case by case that every pair that fits either inequality also fits the other.

Then go over the examples in Question 3. As with Question 1, bring out that there is more than one possible sequence of steps and that there isn't just a single "right" answer.

Homework 3: Simplifying Cookies

This straightforward assignment gives students an opportunity to follow up on today's work using the inequalities from the unit problem.

My Simplest Inequality

In *Homework 2: Investigating Inequalities,* you started with a true inequality involving numbers and explored which operations you could do to both sides that would result in another true inequality.

When inequalities involve variables, we want to know whether the operation produces an **equivalent inequality.** As with equations, two inequalities are called *equivalent* if any number that makes one of them true will also make the other true.

For example, the inequalities $x + 2 < 9$ and $2x + 4 < 18$ are equivalent because, in both cases, the numbers that make them true are precisely the numbers less than 7. For instance, substituting 5 for x makes both true but substituting 10 for x makes both false. (That is, $5 + 2 < 9$ and $2 \cdot 5 + 4 < 18$ are both true, and $10 + 2 < 9$ and $2 \cdot 10 + 4 < 18$ are both false.)

Continued on next page

Part I: One Variable Only

If an inequality has only one variable, you can often find an equivalent inequality that essentially gives the solution. For instance, by subtracting 2 from both sides of the inequality $x + 2 < 9$, you get the equivalent inequality $x < 7$. This tells you that the solutions to $x + 2 < 9$ are the numbers less than 7 (and only those numbers).

1. For each of these inequalities, perform operations to get equivalent inequalities until you obtain one that shows the solution.

   a. $2x + 5 < 8$

   b. $3x - 2 \geq x + 1$

   c. $3x + 7 \leq 5x - 9$

   d. $4 - 2x > 7 + x$

Part II: Two or More Variables

When an inequality has more than one variable, you can't put it into a form that directly describes the solution. But you can often write the inequality in a simpler equivalent form, such as by combining terms.

For example, suppose you start with the inequality

$$9x - 4y - 2 \geq 3x + 10y + 6$$

You can do these steps to get a sequence of simpler equivalent inequalities.

$9x - 2 \geq 3x + 14y + 6$	(adding $4y$ to both sides)
$6x - 2 \geq 14y + 6$	(subtracting $3x$ from both sides)
$6x \geq 14y + 8$	(adding 2 to both sides)

Because all the coefficients in the inequality $6x \geq 14y + 8$ are even, you can do the additional step of dividing both sides by 2, to get $3x \geq 7y + 4$. Each of the inequalities in the sequence is equivalent to the original inequality, but $3x \geq 7y + 4$ seems to be the simplest of them all.

2. a. Find numbers for x and y that fit the inequality $3x \geq 7y + 4$.

   b. Substitute the numbers that you found in Question 2a into the original inequality, $9x - 4y - 2 \geq 3x + 10y + 6$, and verify that they make it true.

Continued on next page

c. Find numbers for x and y that do not fit the inequality $3x \geq 7y + 4$.

d. Substitute the numbers that you found in Question 2c into the original inequality, $9x - 4y - 2 \geq 3x + 10y + 6$, and verify that they make it false.

e. Explain why steps a through d are not enough to prove that the two inequalities are equivalent.

3. For each of the next three inequalities, perform appropriate operations to get simpler equivalent inequalities.

a. $x + 2y > 3x + y + 2$

b. $\frac{x}{2} - y \leq 3x + 1$

c. $0.2y + 1.4x < 10$

Homework 3 Simplifying Cookies

As you have seen, the constraints in the unit problem can be expressed as inequalities using two variables. If you use P to represent the number of dozens of plain cookies and I to represent the number of dozens of iced cookies, one way to write these inequalities is

$P + 0.7I \leq 110$ (for the amount of cookie dough)

$0.4I \leq 32$ (for the amount of icing)

$P + I \leq 140$ (for the amount of oven space)

$0.1P + 0.15I \leq 15$ (for the amount of the Woos' preparation time)

1. Find at least one equivalent inequality for each of the "cookie inequalities" above. If possible, find an equivalent that you think is simpler than the inequality given.
2. For each of the original inequalities, do each of these steps.
   a. Find a number pair for P and I that fits the inequality and a number pair that does not.
   b. Verify that the number pair that fits the inequality also fits any equivalents you found for that inequality.
   c. Verify that the number pair that does not fit the inequality also does not fit any of the equivalents you found for that inequality.

Picturing Cookies

You have turned the bakery's problem into a set of inequalities and a profit expression, but that's just a first step toward understanding the problem. Over the next several days, you will be examining how to represent these inequalities using graphs in a way that gives you, at a glance, a picture of what the Woos' options are.

This page in the student book introduces Days 4 through 7.

Mark Hansen, Jennifer Rodriguez, Karla Viramontes, and Robin LeFevre make a group graph for a cookies inequality.

Picturing Cookies

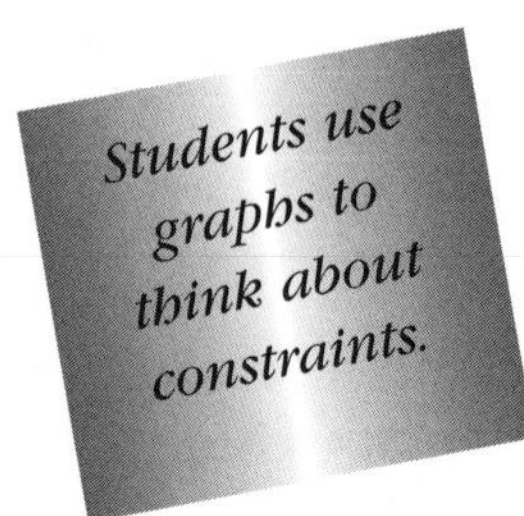

Mathematical Topics

- Simplifying inequalities with two variables
- Discovering the relationship between the graph of a linear inequality and the graph of the related linear equation

Outline of the Day

In Class

1. Discuss *Homework 3: Simplifying Cookies*
   - Bring out that there are many possible equivalents for each inequality
2. *Picturing Cookies—Part I*
   - Students discover how to graph an inequality
   - The activity will be discussed on Day 5

At Home

Homework 4: Inequality Stories

Special Materials Needed

- Grid chart paper
- Rulers
- Colored pencils (in case students forget them)

Discuss With Your Colleagues

Isn't Coloring Dots Too Childish for High School?

Having students coloring in dots on graph paper might seem childish. Why might such an assignment be important for high school students?

1. Discussion of *Homework 3: Simplifying Cookies*

Have various diamond card students present the equivalents they found for each of the cookies inequalities. Try to elicit a variety of responses. For example, students might multiply both sides of the preparation time constraint, $0.1P + 0.15I \leq 15$, by 100 to get

$$10P + 15I \leq 1500$$

Students may need some review work on decimals here, lest they end up with something like $P + 15I \leq 1500$.

Students should see that this inequality is also equivalent to $2P + 3I \leq 300$. There is no "best" equivalent for the inequality, and there is no need to have students look for the greatest common divisor of the coefficients or anything like that. It is often helpful, though, to eliminate decimals. Students also should be encouraged to simplify equations or inequalities when it is easy to do so.

Students should see simplification of inequalities as an issue of convenience rather than one of "right versus wrong." Students who prefer to leave the inequality $0.1P + 0.15I \leq 15$ in this form should be allowed to do so.

For the icing inequality, $0.4I \leq 32$, students should see that, because this involves only one variable, it does have a "simplest" equivalent, namely, $I \leq 80$.

You may want to skip discussion of Question 2, because students will be doing that kind of work in the next activity.

2. *Picturing Cookies—Part I*

In today's activity, *Picturing Cookies—Part I,* students will work from scratch to develop the graph of one or more of the inequalities from the central unit problem. In our experience, even students who seem quite comfortable graphing linear equations or who have graphed inequalities before are often not clear about the ideas in this activity.

> *Comment:* In most algebra texts, students are taught to graph inequalities like these by first graphing the corresponding equality and then shading the appropriate side of that line. This shortcut is definitely *not* the approach that this activity takes. We urge you to insist that students follow the instructions in the activity, even if they already know the shortcut. The experience of actually plotting individual points that fit the inequality and seeing the graph "evolve" will give students a much deeper appreciation of the relationship between the graph of the equation and the graph of the inequality.

"How might you obtain a geometric picture of these constraints?"

To introduce the activity, you can refer the class to the list of constraint inequalities if compiled for the unit problem. Ask students how they might obtain a geometric picture of these constraints. If no one suggests graphing, ask them whether that seems possible and how they might organize such a graph.

• *Choosing axes*

One step in creating a graph is deciding on the axes. For this initial experience with graphing inequalities, we recommend that you have the whole class discuss the issue of choosing which axis will represent which variable.

Bring out that there is no reason why a particular variable should be on the horizontal axis and the other on the vertical axis, because neither variable is more "dependent" or "independent" than the other. Also bring out that it will be easier for students to understand and compare one another's graphs if the class comes to a consensus.

Graphs in this teacher's guide and in the take-home assessment for the unit use the horizontal axis for P and the vertical axis for I, so we recommend that your students do the same.

Note: In some other assignments, such as *Homework 5: Healthy Animals,* each student or group will make an individual choice of axes. This will show students what can happen if there is no common decision about axes.

• *The meaning of "the graph of an inequality"*

"What does it mean to graph an inequality in two variables?"

Ask the class what it means to graph an inequality in two variables. Help students understand that, as with graphing an equation, graphing an inequality means marking all the number pairs that fit the condition. Students may not yet realize that the graph of an inequality is a two-dimensional area rather than a line or a curve, and you should let them discover this through the activity.

• *Doing the activity*

With this introduction, have students work in groups on *Picturing Cookies—Part I.* Give each group two sheets of grid chart paper for use with steps 1 through 3 of Question 1. (One sheet of grid chart paper is for students to experiment on initially; the other is for them to use after they have a sense of what their scales should be. Groups that move on to Question 2 will need more grid chart paper.)

By following steps 1 through 3, students should be able to graph the oven-space inequality. If groups move on to other constraints, emphasize that each constraint is to be treated as a separate problem, on its own set of axes. (In *Picturing Cookies—Part II,* introduced on Day 6, students will look at combining graphs of inequalities.)

Use the suggestions in the subsection "Issues to watch for" as a guideline for how to nudge groups along if they are having difficulty. But be careful not to help students too much. It is important that they eventually feel comfortable with the graphing ideas in this activity, and their understanding will be greater if they develop the ideas on their own.

You may want to let each group produce a single group product, in the form of a poster, rather than have each student prepare a separate written report.

A whole-class discussion of the activity is scheduled for tomorrow. Students need to deal only with the inequality $P + I \leq 140$. Question 2 is for groups that finish ahead of the rest.

- *Issues to watch for*

Be sure students realize that all members of a group should be using the same set of coordinate axes. After each group member finishes checking a number pair, he or she should add the point, in the appropriate color, to the graph. The idea is to accumulate a lot of points quickly, so that students see the graph emerge and can recognize the overall relationship.

You may want to remind students to label their axes appropriately (for example, as "number of dozens of plain cookies" and "number of dozens of iced cookies").

As already noted, we don't want students simply to graph the equation that goes with a constraint (for example, $P + I = 140$ for $P + I \leq 140$) and then shade one side of the line. If students are using the graph of the equation as the basis for their work, urge them to follow the specific directions for the activity. That is, they should simply make up some pairs of numbers for P and I, test them in the inequality, and then color them appropriately.

"What color should (5, 10) be?"

If students are stuck, ask them what color should be used for the number pair $P = 5, I = 10$. Because this pair satisfies the inequality $P + I \leq 140$, they should see that they need to mark the point (5, 10) with the first color. Then they should make up more examples of their own.

Check that they include examples that don't fit the inequality, as well as those that do. For instance, if they try the pair $P = 100$, $I = 100$, they should see that it fails to satisfy the inequality. Therefore, they should mark the point (100, 100) with the second color.

"What color should (40, 100) be?"

As you circulate among groups, you may want to ask them about a pair like (40, 100) in order to bring out that if a number pair fits the equation $P + I = 140$ then it also fits the inequality $P + I \leq 140$.

Finally, you can have groups pool results if some groups are behind.

- *Selecting scales*

"What are the maximum values you will need on each axis?"

After some experimentation, students may find that they can't fit points that don't work within their existing coordinate system. If you see them working with inappropriate scales, you can suggest that they redraw their graphs with new scales, asking them to think about what scales might be appropriate.

They may see that for this inequality, the largest value they will need for either variable for points that satisfy the inequality is 140. However, they should choose their scales so that they can show points that *do not* satisfy the inequality, as well as points that *do*.

You may want to bring up this issue tomorrow as part of the class discussion, asking if any groups had to adjust their scales after starting on the problem.

- *The "big picture"*

Encourage students to keep plotting more and more points, both those that satisfy and those that do not satisfy the inequality. Presumably, they will eventually have an "aha!" and realize that the boundary between *satisfying the inequality* and *not satisfying the inequality* is the line corresponding to the equation $P + I = 140$. This is the "big picture" referred to in step 3.

- *Question 2*

If students get to the other constraints (Question 2), be aware that the "icing inequality" may present special problems because it involves only one variable. You can find suggestions for working with this inequality at the end of tomorrow's discussion of *Picturing Cookies—Part I* (see the subsection "The other constraints").

- *The whole graph of the inequality*

The issue of restricting P and I (that is, not allowing negative or fractional values) may well come up at some point in the context of working on this activity, either during group work or in tomorrow's whole-class discussion. But keep in mind that the goal of this activity is for students to gain insight into graphs of the inequalities. Therefore, if students are focusing on the actual graphs of the inequalities (without reference to the restrictions), we recommend that you defer this issue until Day 6. The section "First Quadrant Only" and subsection "Whole numbers only?" from Day 6 give suggestions on how to introduce and deal with these restrictions in the context of discussion of relevant problems.

There are two scenarios under which you may need to discuss the restriction issue. First, if students raise it on their own. Second, if students limit themselves to positive values (without identifying it as an issue) and therefore do not get the complete graphs. In either case, you can use the relevant discussion from Day 6, as well as comments from Day 2 (see the section "Restrictions on P and I"), for guidance.

Emphasize that the focus of this activity is the graphing, and not how well the inequalities describe the real-world problem. Therefore, at least for now, students should consider all numbers to be "eligible" and should focus on whether they fit the given condition.

Homework 4: Inequality Stories

This assignment will give students more experience relating inequalities to real-world situations.

Picturing Cookies—Part I

By graphing relationships, we can turn symbolic relationships into geometric ones. Because geometric relationships are visual, they are often easier to think about than algebraic statements.

One of the constraints in *How Many of Each Kind?* is that the Woos can make at most 140 dozen cookies altogether (because of oven-space limitations). You can represent this constraint symbolically by the inequality

$$P + I \leq 140$$

where P is the number of dozens of plain cookies and I is the number of dozens of iced cookies.

Choose one color to use for combinations of plain and iced cookies that satisfy the constraint—that is, combinations that total 140 dozen cookies or fewer. Choose a different color for combinations that do not satisfy the constraint—that is, combinations that total more than 140 dozen cookies.

Continued on next page

Some Examples

For instance, what color should you use for the point (20, 50)? In other words, does the combination of 20 dozen plain cookies and 50 dozen iced cookies fit the constraint or not? You can check by substituting 20 for P and 50 for I in the inequality $P + I \leq 140$. Because $20 + 50 \leq 140$ is a true statement, the first color should be used for the point (20, 50).

What about 90 dozen plain cookies and 120 dozen iced cookies? This does not satisfy the constraint, because the statement $90 + 120 \leq 140$ is not true. Therefore, the second color should be used for the point (90, 120).

Your Task

Your task is to plot both types of points and then to describe the graph of the inequality itself. (The graph of the inequality consists of all points that fit the constraint, that is, all points of the first color.)

Steps 1 through 3 in Question 1 give details on what you need to do. Do your final diagram on a sheet of grid chart paper. If you have time, do Question 2, dealing with other constraints.

1. Go through these steps for the oven-space inequality.

   Step 1: Have each group member try many pairs of numbers for the variables, testing whether each pair satisfies the constraint. *On one shared set of coordinate axes,* group members should plot their number pairs using the appropriate color.

   Step 2: Make sure that your group has many points of both colors. After some experimentation, you may need to change the scale on your axes so that you can show both types of points. If necessary, redraw your axes with a new scale and replot the points you have already found.

   Step 3: Continue with Steps 1 and 2, adding points of each type in the appropriate color. Keep going until you get the "big picture," that is, until you are sure what the overall diagram looks like. Include with your final diagram a statement describing the graph of the inequality itself (the points of the first color) and explaining why you think your description is correct.

2. Graph each of the remaining constraints on its own set of axes. Either follow the process described in steps 1 through 3 of Question 1 or use what you learned in Question 1 about the "big picture."

Homework 4 Inequality Stories

You have seen that certain real-world situations can be described using inequalities.

For example, in *How Many of Each Kind?* each dozen plain cookies uses 1 pound of cookie dough and each dozen iced cookies uses 0.7 pounds of cookie dough, but the Woos have only 110 pounds of cookie dough. This limitation can be described by the inequality $P + 0.7I \leq 110$, where P is the number of dozens of plain cookies and I is the number of dozens of iced cookies.

In this assignment, you will look further at the relationship between real-world situations and inequalities.

Part I: Stories to Inequalities

In each of Questions 1 and 2, use variables to write an inequality that describes the situation. Be sure to explain what your variables represent.

1. Rancher Gonzales has built the corral for her horses. Now she's building a pen for her pigs. She isn't so worried about efficiency for the pigs, and she decides to go with a boring old rectangle. The pigs need at least 150 square meters of area. She has to decide what dimensions to use for the pen.

2. Al and Betty want to buy a really fancy spinner that costs $200. They each have some money of their own. Al's parents will contribute $2 for every $1 that Al spends. Betty's grandmother will exactly match Betty's contribution. But even if Al and Betty combine all their own money with these additional funds, they still won't have enough.

Part II: Inequalities to Stories

For each of these inequalities, make up a real-world situation that the inequality describes. Again, be sure to explain what your variables represent.

3. $r < t + 2$

4. $a + b + c \leq 30$

5. $x^2 + y^2 \geq 81$ (If you need a hint, think back to some work you did in *Do Bees Build It Best?*)

DAY 5 Continuing to Picture Cookies

Students see that the graph of a linear inequality is a half plane.

Mathematical Topics

- Expressing situations in terms of inequalities and representing inequalities by situations
- Seeing that the graph of a linear inequality is a half plane

Special Materials Needed

- Colored paper or similar material for acting out *POW 11*

Outline of the Day

In Class

1. Discuss *Homework 4: Inequality Stories*
   - Focus on clear definitions of variables
2. Discuss *Picturing Cookies—Part I* (from Day 4)
   - Focus on the relationship between an inequality and the related equation
   - Bring out that the constraints involve *linear* expressions
   - Introduce the term **half plane**
3. Introduce *Homework 5: Healthy Animals*
   - Get students started on choosing variables and setting up inequalities
4. Introduce *POW 11: A Hat of a Different Color*
   - Have students act out a simple scenario

At Home

Homework 5: Healthy Animals

POW 11: A Hat of a Different Color (due Day 10)

1. Discussion of *Homework 4: Inequality Stories*

You can assign one problem to each group to discuss in detail while you check off the homework (which means there probably will be two groups working on each problem). You might suggest that each group working on Questions 3, 4, or 5 select a couple of possible "stories." Then have the club card members of different groups report to the whole class on the group's discussion.

Be sure that the variables are clearly defined. Question 1 should be straightforward, with an inequality like $LW \geq 150$ coming out of the presentation.

On Question 2, students should come up with something equivalent to $A + 2A + B + B < 200$. (Some students may have trouble with the phrase "contribute $2 for every $1 that Al spends" and mistakenly get $A + 0.5A + B + B < 200$ instead.)

For Questions 3 through 5, students will probably have a wide variety of responses. For example, stories for Question 5 might involve distances, the Pythagorean theorem, or area.

2. Discussion of *Picturing Cookies—Part I*

"What do the different colors represent?"

Different groups will probably grasp the idea of this activity at different rates. As soon as all groups are done with steps 1 through 3 for the oven-space constraint, you can have groups present their final diagrams and then bring the class together to discuss their work. Be sure to have groups making presentations relate the different colors to the idea of *satisfying the inequality* and *not satisfying the inequality.*

The main goal of this discussion is to articulate the "big picture" referred to in step 3, which is the relationship between the graph of the inequality and the graph of the corresponding equation. That is, be sure students see that the graph of the equation forms the boundary between points that satisfy the inequality and points that do not. Students should state this principle in their own words.

Note: A group presenting its work may focus on its finished product, having understood clearly the relationship between the equation and the inequality. Other groups may still be unclear. Therefore, it is important to encourage the class to question the presenters until the explanation is clear to everyone.

- ***Why is $P + I = 140$ the boundary?***

Ask if someone can explain why all the solutions to the inequality are on one side of the graph of $P + I = 140$. A student might explain that if you move to the left or down from the line $P + I = 140$, either P or I will decrease, so the value of $P + I$ will go down. If this value was equal to 140 *on the line,* then this value will be *less than* 140 to the lower left of the line.

Note: We often get lazy about distinguishing between an equation and its graph. For example, when we say, "the line $P + I = 140$," we really mean "the line that is the graph of the equation $P + I = 140$." You should point out this verbal shorthand to students. The interconnectedness of the ideas of equation and graph can never be overemphasized.

Introduce the term **half plane** to describe the set of points that fit the inequality. This may need some explanation, because the term *plane* may be unfamiliar to some students.

In particular, be sure students realize that a plane is infinite in extent. They might describe a plane as "an infinite flat surface." You need not get technical about the geometry here. You might point out that a half plane has somewhat the same relationship to a plane that a ray does to a line. You may need to remind students, in the context of this discussion, that a *line* is also infinite in extent.

Note: The term *half plane* is sometimes defined to exclude the line itself. Be sure students realize that the graph of the inequality $P + I \leq 140$ includes the points on the line $P + I = 140$.

Students will learn more about planes in the Year 3 unit *Meadows or Malls?* which extends the ideas in this unit to several variables. In that unit, students will work with, and then graph, linear inequalities in three variables, and then go on to situations with more variables.

- ***Why is the boundary a straight line?***

"Why is the graph of P + I = 140 a straight line?"

Once students are clear on why the graph of the equation $P + I = 140$ forms the boundary between the two colors, you can review the concept of a *linear expression* by asking why the graph of this equation is a straight line. This need not be a technical discussion, but simply a review of ideas from *Solve It!* connecting the algebraic form of expressions such as $P + I$ with the geometry.

Bring out that all of the algebraic expressions in the unit problem, both those in the constraints and the profit expression, are linear. Tell students that an inequality involving linear expressions is called, appropriately, a **linear inequality.**

- ***The other constraints***

After this discussion, you can have groups that finished other constraints present their results. You may decide that you don't need to discuss all the constraints but be sure the class at least sees each graph. Do take some time to consider the constraint $0.4I \leq 32$ (or an equivalent such as $4I \leq 320$ or $I \leq 80$), because the absence of the variable P may present a special difficulty. As a hint, you can ask students how they graph an *equality* with only one variable, such as $0.4I = 32$. Try to elicit the idea that P can be anything.

You may want to suggest rewriting this inequality as $0P + 0.4I \leq 32$. You can also refer back to the context of the problem, asking, "If the only restriction on the Woos was the amount of icing they had, how many dozens of plain and of iced cookies could be made?" Students should see that this constraint by itself does not limit the number of dozens of plain cookies at all.

3. Introduction to *Homework 5: Healthy Animals*

You may need to take a few minutes to introduce this homework, primarily to help students choose variables for expressing their inequalities. Their earlier work in setting up the inequalities for the Woos will help you decide how much of this to do with them in class.

They should have enough understanding from today's class activity to be able to do the graphing itself (Question 3) on their own. Don't bring up the issue of which variable to use for which axis, because that will make for an interesting discussion tomorrow. If students ask about this, tell them to make their own choices.

> *Note:* Unlike the cookie problem, this new problem has no expression to be maximized. Students will work further with this situation in the activity *Feasible Diets* on Day 7.

4. Introduction to *POW 11: A Hat of a Different Color*

> *Note:* You will probably need about ten minutes to introduce *POW 11: A Hat of a Different Color.* If necessary, you can delay its introduction until Day 6 and simply have students read the problem tonight.

The most important element of *POW 11: A Hat of a Different Color* is the explanation students give for the fact that Carletta was *sure* what color hat she had on. Their explanation has to take into account all possibilities.

You can help students understand the situation by having them act out a similar but simpler situation. We provide two scenarios here, using one red hat and two blue hats. (You can use pieces of colored paper to represent the hats.) We suggest that at this point, you have students act out just the first of the scenarios. You may want to check in with students in a couple of days and see how they are doing. At that point, if they seem stumped, you can work through the second scenario. (A reminder to do this is included at the beginning of Day 7.)

• *The first scenario*

Begin by showing the three hats to the whole class. Then choose two students and have them close their eyes. Put the red hat on the head of one student and one of the blue hats on the head of the other student; hide the other blue hat.

While the two students still have their eyes closed, ask if either of them knows *with certainty* what color hat is on her or his own head. It should be clear to everyone in the class that neither of them can figure this out.

To the student with the blue hat: "Open your eyes and look at the other student's hat. Can you tell now what color you have?"

Then ask the student with the blue hat to open her or his eyes, look at the other student's hat, and try to determine the color of her or his own hat. This student will see a red hat and should be able to deduce that her or his own hat must be blue, because there is only one red hat. (If this student doesn't figure this out, ask if anyone else can help. Someone should be able to explain the reasoning.)

• *The second scenario*

Have two different students close their eyes. This time put blue hats on both students, and again ask them if they can tell, without opening their eyes, what color their own hat is.

After getting presumably the same negative answer as before, have one of the two students open his or her eyes, and ask this student to try to determine the color of her or his own hat. This time the student sees a blue hat. Because there is both a red hat and a blue hat still unaccounted for, this student can't tell.

Finally, ask the second student what he or she can deduce, *without opening his or her eyes!* The key insight is that if the first student had seen a red hat, that student would have known that he or she had a blue hat. (Your students should see this from the first scenario.) Therefore, the second student must not have a red hat and so must have a blue hat.

If the second student doesn't articulate this reasoning, ask if anyone else in the class can help out. Emphasize that this "outsider" must provide an explanation that doesn't involve actually seeing the second student's hat.

You may wish to leave this unresolved if the class doesn't see it clearly.

• *Groups for POWs*

You might want to have students work in groups on this POW, with the suggestion that they develop a skit or some other way to act out portions of their presentations.

This POW is scheduled for discussion on Day 10.

Homework 5 Healthy Animals

Curtis is concerned about the diet he is feeding his pet. A nutritionist has recommended that the pet's diet include at least 30 grams of protein and at least 16 grams of fat per day.

Curtis has two types of foods available—Food A and Food B. Each ounce of Food A supplies 2 grams of protein and 4 grams of fat, while each ounce of Food B supplies 6 grams of protein and 2 grams of fat. Curtis's pet should not eat a total of more than 12 ounces of food per day.

Curtis would like to vary the diet for his pet within these requirements, and so he needs to know what his options are.

1. Choose variables to represent the amount of each type of food Curtis will include in the daily diet. State clearly what the variables represent.
2. Use your variables to write inequalities to describe the constraints of the problem.
3. Choose one of your constraints. Draw a graph that shows which combinations of Food A and Food B satisfy that constraint. Be sure to label your axes and show their scales.

POW 11 *A Hat of a Different Color*

Once upon a time, many years ago and very far away, there lived a wise high school teacher, whose students were always complaining noisily that they had too much homework (and too many POWs, too!).

The wise teacher offered the three noisiest students a deal. He showed them that he had two red hats and three blue hats. The deal worked like this:

> The three students would close their eyes, and while their eyes were closed, the teacher would put a hat on each of their heads (and hide the other two hats).
>
> Then, one at a time, the students would open their eyes, look at the other two students' heads, and try to determine which color hat was on their own head. At a given student's turn, that student could either guess what color hat he or she had or "pass."

Continued on next page

While the first student's eyes were open and that student was still deciding what to do, the other two students kept their eyes closed.

Once the first student either guessed or passed, then the second student could open his or her eyes and either guess or pass. (The eyes of the third student had to remain shut.)

When the second student was finished, then the third student could open his or her eyes and either guess or pass.

Any student who guessed correctly would have no POWs to do the rest of the semester. But any student who guessed wrong would not only have to do the POWs but also help grade everyone else's work. If a student decided to pass, then the work load would stay as usual.

The students drew numbers to see who would go first. Then they closed their eyes, and the wise teacher put a hat on each one's head and hid the remaining two hats.

Arturo, who was first, opened his eyes, looked at the others' heads, and said he wanted to pass. He couldn't tell for sure, and he didn't want to guess in case he was wrong.

Next, Belicia opened her eyes and looked at the others' heads. She also thought about the fact that Arturo had said he couldn't tell. Then she said she didn't want to risk it either. She couldn't tell for sure.

Carletta was third. She just sat there with her eyes still closed tightly and a big grin on her face. "I *know* what color hat I have on," she said. And she gave the right answer.

Your POW is to figure out what color hat Carletta had on and how she knew for sure. The most important part of your POW write-up will be your explanation of how she knew for sure.

Reminder: Carletta didn't even look! You should also know that all three students were extremely smart, and if there was a way for them to figure it out, they would be sure to do so.

Write-up

1. *Problem Statement*
2. *Process*
3. *Solution:* Explain how you know for sure what color hat Carletta had on.
4. *Evaluation*
5. *Self-assessment*

Completing the Cookie Picture

Students put the cookie constraints together in a single graph.

Mathematical Topics

- Expressing constraints stated in words as inequalities using algebra
- Determining which side of a line is the graph of a linear inequality
- Restricting variables to the first quadrant
- Combining graphs of linear inequalities

Outline of the Day

In Class

1. Discuss *Homework 5: Healthy Animals*
   - Define variables, label the axes, and develop inequalities
   - Discuss the issue of "which side of the line?"
2. Discuss limiting consideration to the first quadrant (if not discussed previously)
   - Introduce new constraints to restrict the variables to nonnegative values
3. *Picturing Cookies—Part II*
   - Students create a composite graph for the inequalities from the unit problem
   - Discuss the issue of whether to restrict variables to whole-number values (if not discussed previously)
   - The activity will be discussed on Day 7

At Home

Homework 6: What's My Inequality?

1. Discussion of *Homework 5: Healthy Animals*

Have students discuss their solutions in their groups. Tell them that they may have done different things in the problem and that they should try to understand one another's work. As you circulate, try to find students who put the amount of Food A on the vertical axis. (Most students will likely put the amount of Food A on the horizontal axis. What you want is to find examples of both approaches so students can see the impact of choice of axes.)

"What differences did you notice in your work?"

Before looking at specific questions from the homework, ask the class what differences they noticed in their work on the assignment. They may mention that the variables could have different names, that the axes could be labeled differently, and that the scales might vary.

• *Choosing variables and writing inequalities*

"Let's all use the same variables. What variables shall we use? What do they represent?"

As a class, decide on variables and what they represent. We will use a and b for the number of ounces of Foods A and B, respectively, and put a on the horizontal axis and b on the vertical axis.

Let spade card students from several groups give inequalities they developed for the problem. If necessary, have them restate their results using the variables decided on as a class. They will probably come up with these inequalities (or equivalents):

$2a + 6b \geq 30$ (to guarantee enough protein)

$4a + 2b \geq 16$ (to guarantee enough fat)

$a + b \leq 12$ (to limit total intake)

You may want to bring out that, as with the inequalities from the unit problem, these inequalities involve linear expressions.

• *Graphs of the inequalities*

Once these inequalities are agreed on, let spade card students from other groups present the graphs of these inequalities. If you found students with the amount of Food A on the vertical axis, try to include one such presenter for each inequality.

Have students comment on the differences among the various graphs for each inequality. Such differences may come both from different labeling of axes and from different scales, so be alert to these variations and the confusion they may cause for students.

Because this problem will be revisited tomorrow in the activity *Feasible Diets,* we recommend that you agree now on labels and scales for the axes. But be sure to point out as well that there is no "right" labeling or scaling for the graphs from the homework.

Students may want to save their work so that they can refer to it when they work on *Feasible Diets.*

- ***Which side of the line?***

Notice that for the first two of the inequalities listed previously, the graph is the area above and to the right of the graph of the corresponding equation. By contrast, all the graphs of the inequalities in the unit problem were below and to the left of the corresponding equation. So this is a good time to ask students how to tell which side of the line they want. (Clarify that the line itself is included as well, unless the inequality is strict.)

"How do you tell which side of the line you want?"

Perhaps the simplest method is to pick a point that is off the line and see if it satisfies the inequality. For example, to see if the inequality $2a + 6b \geq 30$ represents the half plane above the line $2a + 6b = 30$ or the half plane below that line (as well as the line itself), consider an easy point such as (0, 0). This point is *below* the line and *does not* satisfy the inequality $2a + 6b \geq 30$. Therefore, the points that *do* satisfy the inequality form the half plane *above* the line.

Similarly, the point (2, 0) is below the line $a + b = 12$ and satisfies the inequality $a + b \leq 12$. Therefore, the graph of the inequality $a + b \leq 12$ must be the half plane *below* the line $a + b = 12$.

Note: It may seem obvious that "≥" means *above* and "≤" means *below.* But the inequality $a + b \leq 12$ is the same as $12 \geq a + b$, so the direction of the inequality sign doesn't by itself determine which side of the line represents the graph of the inequality.

Also, it is not obvious what to do when there are variables on both sides of the inequality, as in an inequality like $2x \geq y - 3$.

The question of which side of a line to use is confusing to many students. Having them try to learn a system of rules (for instance, involving signs of coefficients) may lead to unnecessary anxiety about graphing inequalities. Instead, we recommend that they simply test an example or two to decide which side of the line they want in a given situation. This system will help them develop confidence in their own skills.

2. First Quadrant Only

If it hasn't yet been discussed, this is a good time to bring up the fact that it often doesn't make sense in a real-world problem for variables to be negative. In *Healthy Animals,* both a and b must be nonnegative.

It may be especially helpful to discuss this fact in relation to this problem, because you can do it without raising the other restriction issue mentioned on Day 2, namely, limiting the values of the variables to integers. In *Healthy Animals,* all nonnegative numbers make sense.

"Is (-1, 11) a possible choice for Curtis's pet's diet? Doesn't it fit the constraints?"

Here is one way to bring up this idea. While looking at the graph from one of the inequalities, such as $2a + 6b \geq 30$, ask whether a specific point, such as (-1, 11), is a possible choice for Curtis's pet's diet. If students say no, argue that the coordinates do satisfy the inequality, and in fact satisfy all three inequalities, and therefore the point should be an option for Curtis.

Presumably, students will claim this doesn't make sense—that is, that an amount of food cannot be negative. Emphasize that this condition does not come from any of the three constraint inequalities presented previously, but is inherent in the meaning of the variables in the problem context.

You can suggest to students that they need to distinguish between the problem of finding possible diets for Curtis's pet and the problem of graphing the particular inequality. If they are doing the latter, then the point (-1, 11) should be included, because it satisfies the inequality. But, of course, they don't want to allow "-1 ounces of Food A, 11 ounces of Food B" as an acceptable diet for Curtis's pet.

"How can you impose this nonnegativity condition on the overall problem?"

Then ask how they could impose this nonnegativity condition on the overall problem. As a hint, you can suggest that they put additional inequalities into the problem. They should see that they can do this by adding two additional constraints, namely, $a \geq 0$ and $b \geq 0$.

In anticipation of the next activity, you may want to suggest that students make a similar adjustment to their constraint list for the unit problem (posted on Day 2). Or you may prefer to see if they make this adjustment on their own as they work on the activity.

3. *Picturing Cookies—Part II*

In *Picturing Cookies—Part I,* students learned how to graph each of the inequalities from the unit problem. Today, in *Picturing Cookies—Part II,* they will look at how to combine the graphs of the inequalities. They will discuss this activity tomorrow.

> *Note:* You will probably want all students to make their own graphs so that they can refer to them later and include them in their portfolios.

The major new idea introduced in this activity is the principle that the set of points that simultaneously satisfy more than one linear inequality is the intersection or overlap of the individual half planes for the separate inequalities.

Students have encountered the term *intersection* before, in the context of graphs of equations (for example, in the Year 1 unit *The Overland Trail*). There, as here, it refers to the set of points that lie on the graphs for two or more different conditions. The intersection points represent solutions that are common to all the conditions.

In the case of two linear equations and their graphs, the intersection simply refers to the point where the lines cross. This special case is important later in this unit.

Picturing Cookies—Part II takes students step-by-step through the analogous process, both for the sake of review and because the case of inequalities is more complex and more intimidating than that of equalities.

• *Whole numbers only?*

As mentioned on Day 2, the values for P and I in the answer to the Woos' problem should probably be whole numbers (or at least multiples of $\frac{1}{12}$). If this issue hasn't come up before, it should definitely be introduced in connection with this activity. You can raise it with individual groups as they work or wait until tomorrow's whole-class discussion to see if students bring it up.

"Does (52.35, 28.71) fit the constraints? Could these be the numbers of dozens of each kind of cookie?"

• *Suggestions for discussing the issue*

If you need to bring this issue up yourself, you can simply name a point that fits all the constraints, such as (52.35, 28.71), and ask if the Woos could make that many dozens of plain and iced cookies. Students should recognize that this doesn't make sense.

No matter how the issue comes up, you can point out that in graphs like these, with values of 100 and more for P and I, it would be very awkward to try to graph individually each of the points with whole-number coordinates.

Suggest that to start, students work with the complete graphs of the inequalities, including points whose coordinates are not whole numbers. These graphs can still provide us with a useful picture, even though they contain extraneous points.

In other words, tell students that for now, they should graph only the inequalities themselves; they can worry later on about how fully the inequalities model the real-world situation. If it turns out that the point that gives the maximum profit has coordinates that are "ineligible," they will have to decide where to go from there.

In this problem, the maximum profit does occur at a point with whole-number coordinates. In problems for which this is not the case, the task of finding the optimal whole-number solution can be quite complicated. (Taking the whole-number point closest to the optimal overall solution does not always give the optimal whole-number solution.)

Homework 6: What's My Inequality?

In *Picturing Cookies—Part I,* students saw that the graph of a linear inequality is a half plane. In tonight's homework, they work in the other direction, starting with the graph and developing the corresponding algebraic statement.

In Part I, they are given a straight line and asked to find the corresponding linear equation. In Part II, they are given a half plane and the equation of its boundary, and asked to find the linear inequality for the half plane. This assignment should deepen their understanding of the relationship between algebra and graphs in general, and of the graphs of inequalities in particular.

Students Keenzia Budd and Kolin Bonet compare cookies to see which one is bigger.

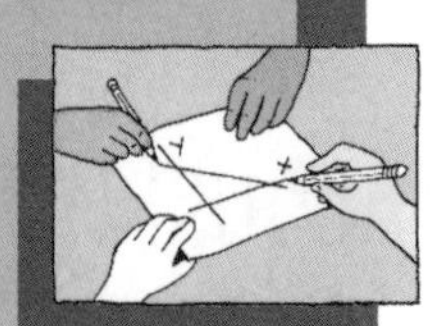

Picturing Cookies—Part II

You have already graphed each of the constraints from the unit problem on its own set of axes. Each graph gave you a picture of what that constraint means.

Now you need to see how to combine these constraints to get one picture of all of them together.

1. Begin with one of the constraints that you worked on before. Using a colored pencil, color the set of points that satisfy this constraint. (*Note:* Unlike your work on *Picturing Cookies—Part I,* you should *not* color the points that fail to satisfy the constraint.)

2. Now choose a second constraint from the problem.

   a. On the *same set of axes,* but using a *different color,* color the set of points that satisfy this new constraint.

   b. Using your work so far, identify those points that satisfy *both* your new constraint and the constraint used in Question 1.

3. Continue with the other constraints, using the same set of axes. Use a new color for each new constraint.

   a. Color the set of points that satisfy each new constraint.

   b. After graphing each new constraint, identify those points that satisfy all the constraints graphed so far.

4. When you have finished graphing all the constraints, look at your overall work. Make a single new graph that shows the set of all those points that represent possible combinations of the two types of cookies the Woos can make. In your graph, show all of the lines that come from the constraints, labeled with their equations.

Homework 6 What's My Inequality?

Graphs of inequalities play an important role in understanding some problem situations. In *Picturing Cookies—Part I,* you started from an algebraic statement—a linear inequality from the unit problem—and saw that its graph was a half plane.

In this assignment, you go from graphs back to algebra. In Part I, you are given graphs that are straight lines, and your task is to find the corresponding linear equations. In Part II, you are given the equation for a straight line, and your task is to find the inequality corresponding to the half plane on one side of that line.

Part I: Find the Equation

For each of the straight lines in graphs 1 through 4, write a linear equation whose graph is that straight line.

Also describe in words the process by which you found the equation.

1.

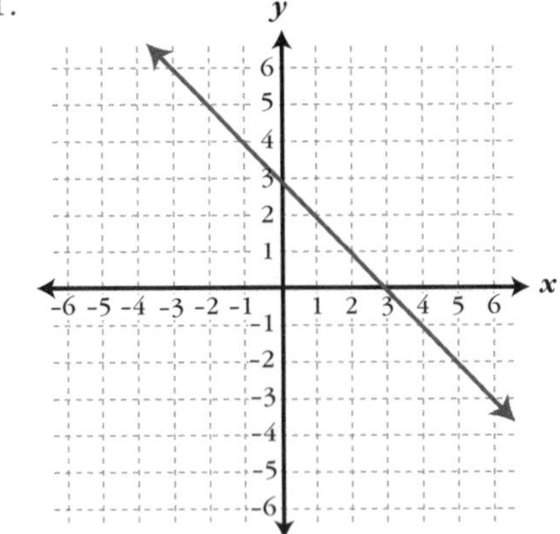

Continued on next page

2.

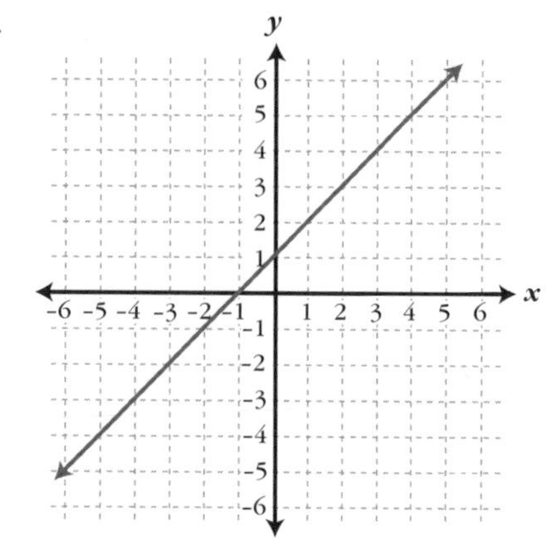

3.

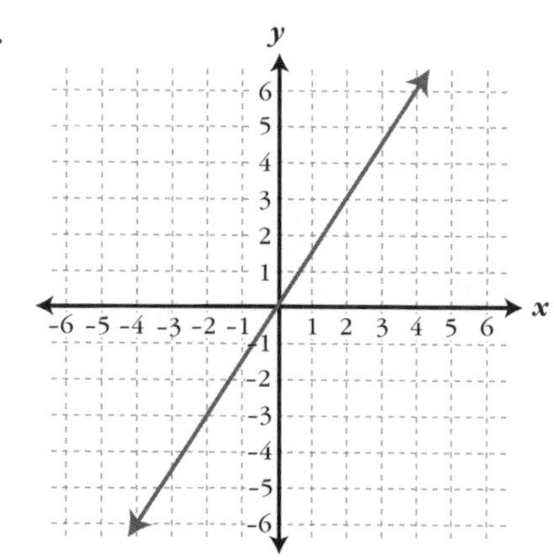

4.

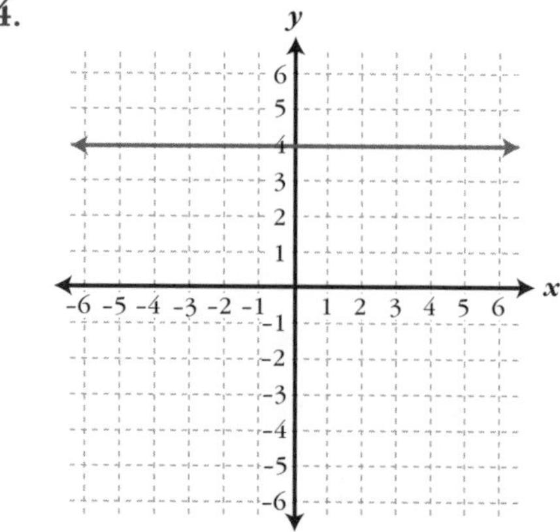

Continued on next page

Part II: Find the Inequality

The shaded area in each of graphs 5 through 8 represents a half plane. (You should imagine that the shaded area continues indefinitely, including all points on the shaded side of the given line.) In each case, you are given an equation for the straight line that forms the boundary of the half plane. Your task is to find a linear inequality whose graph is the half plane itself.

Note: If the boundary is shown as a dashed line, it is not considered part of the shaded area. If the boundary is shown as a solid line, it is considered part of the shaded area. This is a common convention, similar to the convention of open and filled-in circles used in *Homework 2: Investigating Inequalities* for graphs in one variable.

5.

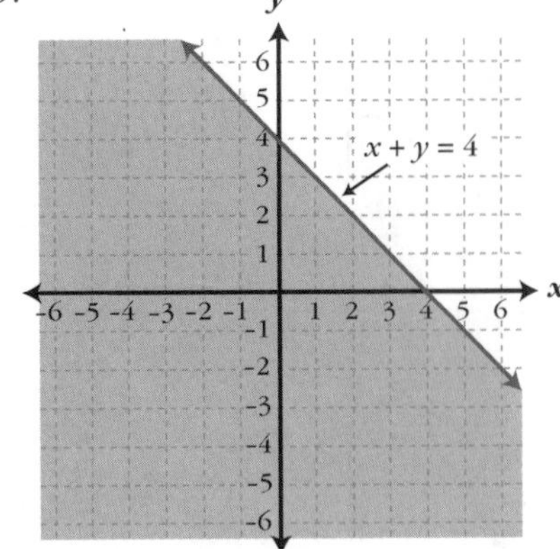

6.

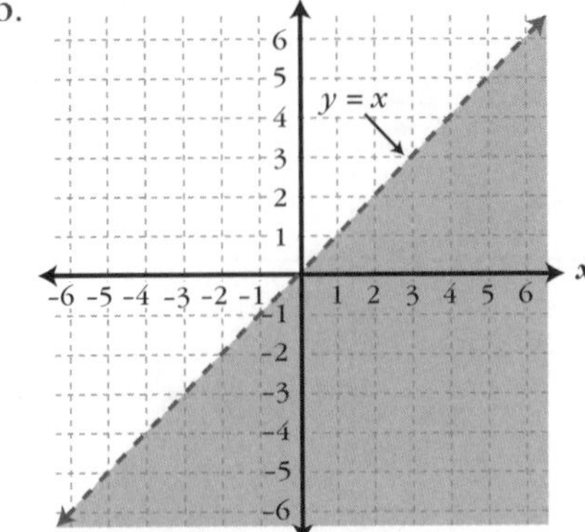

Continued on next page

7.

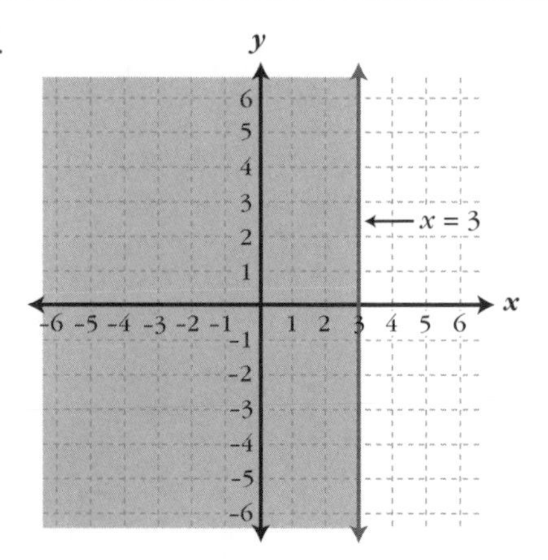

8.

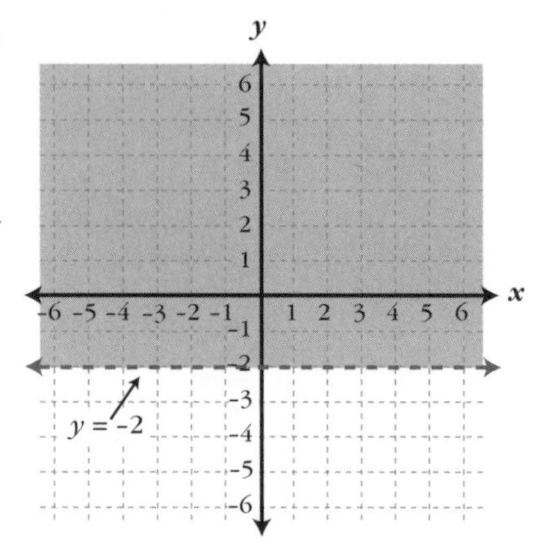

The Cookie Region

The concept of a feasible region is introduced, and students apply it to a new problem.

Mathematical Topics

- Finding the equation for a straight-line graph
- Finding inequalities to describe half planes
- Combining graphs of linear inequalities
- Introducing the concept of a feasible region
- Finding feasible regions

Outline of the Day

In Class

1. Check on students' progress on *POW 11: A Hat of a Different Color*
2. Discuss *Homework 6: What's My Inequality?*
   - For Part II, focus on
     - ✔ the role of the equation as the boundary
     - ✔ the distinction between strict and nonstrict inequalities
     - ✔ determining the direction of the inequality
3. Discuss *Picturing Cookies—Part II* (from Day 6)
   - Introduce the terms **feasible point** and **feasible region**
4. *Feasible Diets*
   - Students apply the concept of a feasible region to the situation from *Homework 5: Healthy Animals*
   - No whole-class discussion is needed for this activity

At Home

Homework 7: Picturing Pictures

Discuss With Your Colleagues

How Good Should Students Get at This?

In *Homework 6: What's My Inequality?* students were asked to find equations or inequalities to fit certain graphs. What is the purpose of such activities? How much mastery do you think students should have of these skills?

1. Progress Check on *POW 11: A Hat of a Different Color*

This is a good time to check how students are doing on *POW 11:A Hat of a Different Color.* Based on their response, you may want to act out the second scenario described in the POW introduction on Day 5.

As a simpler hint, you can emphasize that Carletta's ability to figure out the color of her hat depends on her faith in Arturo's and Belicia's reasoning ability. Carletta must use the fact that they can't figure out the color of their own hats, which would have no significance if they were not both smart students.

2. Discussion of *Homework 6: What's My Inequality?*

You can ask different groups to each prepare a presentation on one of the homework problems. Have heart card students make the presentations.

"Did anyone have a different way to find the equation?"

On Part I, insist that presenters describe how they developed the particular linear equations. Their methods will probably be rather informal and ad hoc, but the class will benefit from seeing a variety of approaches. Whatever strategy a group presents, ask whether other students used other strategies.

Use Part II to review the principle that the line itself forms the boundary of the graph of the inequality and to review the distinction between strict and nonstrict inequalities. This is also an opportunity for continued work on how to determine the direction of the inequality. You can review the technique of making this determination by picking a point not on the boundary as a test case.

3. Discussion of *Picturing Cookies—Part II*

Ask for one or two volunteers to present their combined graphs from *Picturing Cookies—Part II.* Insist that they explain how they got their region, going through the inequalities one at a time.

"Where did that inequality come from?"
"What does that line represent?"
"Why do we want the points on this side of the line rather than the other side?"
"Are points on the line part of the region?"

During the presentations, you can ask the rest of the class questions like these to check that they are understanding what is going on:

- Where did that inequality come from?
- What does that line represent?
- Why do we want the points on this side of the line rather than the other side?
- Are points on the line part of the region?

Be sure the entire process gets tied together at the end. That is, make sure students don't get so involved in the details of individual inequalities that they forget the connection of the graphs to the unit problem.

"What does the final graph tell you about the cookie problem?"

They should be able to articulate that the points that have been colored every time represent the Woos' choices. Be sure that the "first quadrant only" issue has been dealt with (see Day 6), by including the specific constraints $P \geq 0$ and $I \geq 0$.

• *Feasible region and feasible points*

You should play up the importance of the final graph (Question 4), bringing out that this diagram combines much of the information from the problem into a single picture. (The picture does not include information about cost and selling price.)

"How does seeing this combined graph compare to reading the verbal description of the situation?"

Have students look back at the original problem from Day 1, *How Many of Each Kind?* to see what an achievement it is to represent so much information so simply.

You may want to ask students what *they* would call this region. Whatever their response, tell them that in standard mathematics terminology, this set of points is called the **feasible region** for the set of inequalities, and individual points in the region are called **feasible points.** (This is the first of several feasible regions that students will draw.)

You may also want to comment on the everyday use of the word *feasible* to mean "possible" or "capable of being done."

Note: You need not worry about whether students get the coordinates of the points of intersection at this time. If they found them, fine, but if not, that can wait until later in the unit. Students will return to the unit problem on Day 19, at which time they might make a fairly fresh start. Students will be finding points of intersection in connection with other situations (such as *Profitable Pictures,* discussed on Day 9, and *Homework 12: Rock 'n' Rap*). The general issue of points of intersection is the focus of the activity *Get the Point,* which students begin on Day 15.

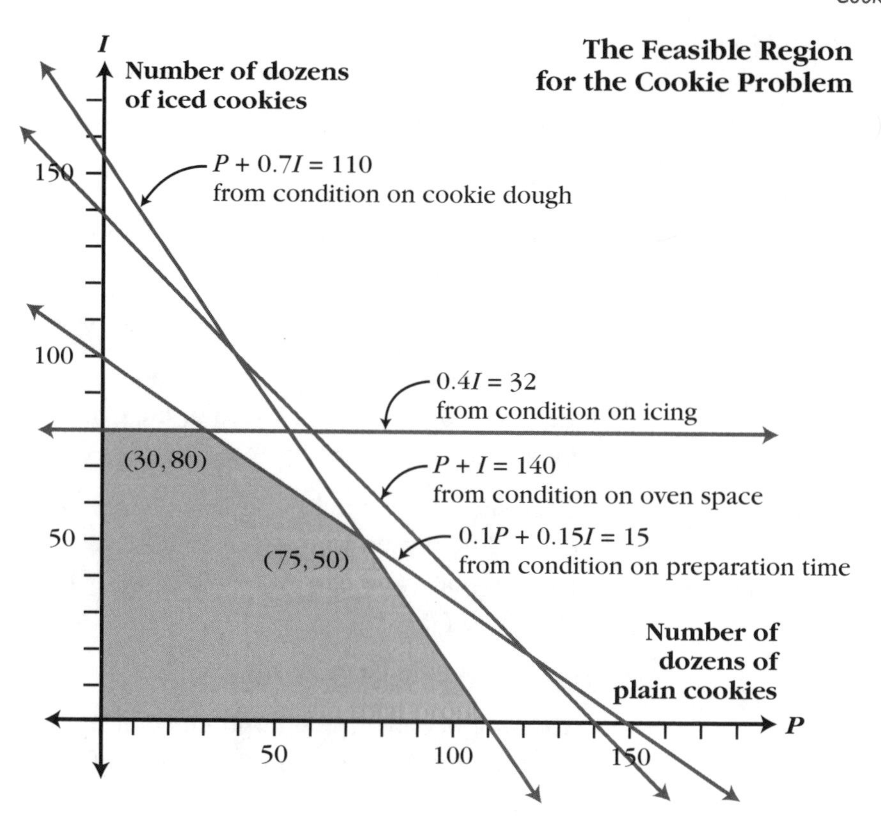

Note: The shaded area in the graph here shows the feasible region for the cookie problem.

- *An extraneous constraint*

"What does it mean that the oven-space line misses the feasible region?"

Bring out that the line $P + I = 140$, which comes from the limitation on the amount of oven space available, misses the region completely, and ask students what this fact means.

> One way of expressing its significance is that, because of the other constraints, it wouldn't help the Woos to have an unlimited supply of oven space. Another way of saying this is that, because the Woos must satisfy all the other constraints, they can't make use of all of their available oven space.

4. *Feasible Diets*

To follow up on this introduction to the concept of feasible region, have students work in groups on the activity *Feasible Diets,* in which they draw the feasible region for the pet-diet problem *Homework 5: Healthy Animals*. (*Reminder:* They graphed the individual inequalities in that assignment and the subsequent discussion, but did not put them together as they are asked to do here.)

We suggest that you have each group prepare a poster showing the feasible region. Use your judgment about whether this needs to be discussed as a

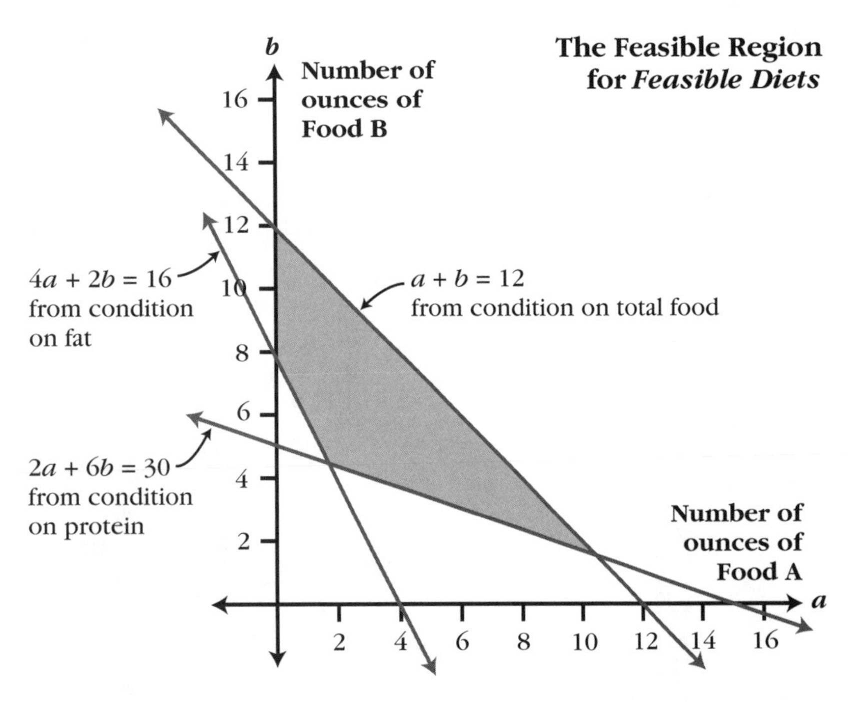

The feasible region for this problem is the shaded area in the graph shown here. In drawing the feasible region, students will need to take into account that the variables must be nonnegative, probably using the inequalities $a \geq 0$ and $b \geq 0$ introduced on Day 6. Without these constraints, the feasible region would continue across the vertical axis to the intersection in the second quadrant of the two lines $4a + 2b = 16$ and $a + b = 12$.

"Should (-1, 11) be in the feasible region? Is it a possible solution to the problem?"

If you see students making the mistake of going outside the first quadrant, you can identify a point that fits the original three inequalities, such as (-1, 11), and ask whether this should be in the feasible region. If needed, follow up by asking if this is a possible solution to the problem.

Reminder: There is no function to be maximized or minimized in this problem.

Homework 7: Picturing Pictures

Tonight's homework introduces a new problem and reinforces the graphing work done so far. It returns to the issue of maximizing profit, which has not been considered since Day 2.

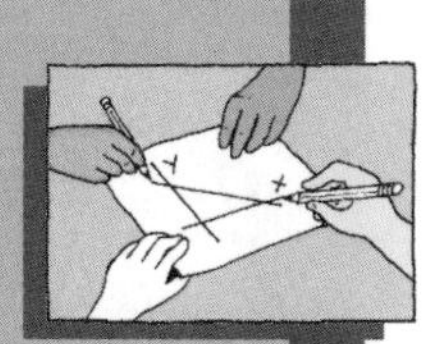

Feasible Diets

You have graphed the individual constraints from *Homework 5: Healthy Animals.* Now your task is to draw the feasible region for that problem.

Here are the key facts.

- Curtis's pet needs at least 30 grams of protein.
- Curtis's pet needs at least 16 grams of fat.
- Each ounce of Food A supplies 2 grams of protein and 4 grams of fat.
- Each ounce of Food B supplies 6 grams of protein and 2 grams of fat.
- Curtis's pet should eat a total of no more than 12 ounces of food per day.

Be sure to identify your variables, label your axes, and show the scales on the axes.

Homework 7 Picturing Pictures

Hassan is an artist who specializes in geometric designs. He is trying to get ready for a street fair next month.

Hassan paints both watercolors and pastels. Each type of picture takes him about the same amount of time to paint. He figures he has time to do a total of at most 16 pictures.

The materials for each pastel will cost him $5, and the materials for each watercolor will cost him $15. He has $180 to spend on materials. He makes a profit of $40 on each pastel and a profit of $100 on each watercolor.

1. Express Hassan's constraints as inequalities, using p to represent the number of pastels he does and w to represent the number of watercolors.
2. Make a graph that shows Hassan's feasible region. In other words, the graph should show all the combinations of watercolors and pastels that satisfy his constraints.
3. For at least five points on your graph, find the profit that Hassan would make for that combination.
4. Write an algebraic expression to represent Hassan's profit in terms of p and w.

Using the Feasible Region

This page in the student book introduces Days 8 through 14.

As you have seen, the collection of inequalities that describes the central problem of the unit can be represented geometrically as the *feasible region*. But how do you use this region to solve the problem? How do you determine which point in the region will maximize the Woos' profit?

In the next portion of the unit, you will look at several problems similar to the bakery one and examine how geometry can help you find the maximum or minimum value of a linear expression on a feasible region.

Myrna Yuson, Daniel Escamilla, and Guru Probhakara proudly display their feasible region for "Rock 'n' Rap."

Profitable Pictures

Students begin to examine the relationship between the profit function and the feasible region.

Mathematical Topics

- Expressing constraints symbolically
- Finding a feasible region
- Finding the value of a profit function for some points in a feasible region
- Exploring the set of points that yield a given profit
- Trying to maximize a profit function

Outline of the Day

In Class

1. Discuss *Homework 7: Picturing Pictures*
   - Post a graph of the feasible region for later use
   - Identify the profit function
2. *Profitable Pictures*
   - Students find combinations that yield a given profit
   - The activity will be discussed on Day 9

At Home

Homework 8: Curtis and Hassan Make Choices

Special Materials Needed

- Transparency of the feasible region for *Homework 7: Picturing Pictures* (see Appendix B)

1. Discussion of *Homework 7: Picturing Pictures*

As a class, decide which variable will go on which axis. We will place p on the horizontal axis and w on the vertical axis, but you should inform students that either choice is correct. Ask the class how to represent the two constraints symbolically, perhaps choosing students at random. They should get inequalities equivalent to these:

$p + w \leq 16$ (for the number of pictures)

$5p + 15w \leq 180$ (for the money available for materials)

Students may remember to include the inequalities $p \geq 0$ and $w \geq 0$. If not, you can bring this up as they sketch the feasible region, as described below.

- *Question 2: The feasible region*

Ask for a volunteer to present the development of the feasible region, and post the graph for further use today and tomorrow.

> If your students have a shaky grasp of the connection between inequalities and their graphs, you should use this as another occasion to review the process. Have students offer specific points that fit the constraints, and then have the class check that they satisfy all of the constraints (including $p \geq 0$ and $w \geq 0$). Continue plotting points until the connection between the developing graph and the equations $p + w = 16$ and $5p + 15w = 180$ becomes clear.

The feasible region for *Homework 7: Picturing Pictures* should look like the shaded area of the graph shown here. You will probably want a transparency of this region (in addition to the posted graph) for tomorrow's discussion of the next activity, *Profitable Pictures*. A large diagram for making this transparency is included in Appendix B.

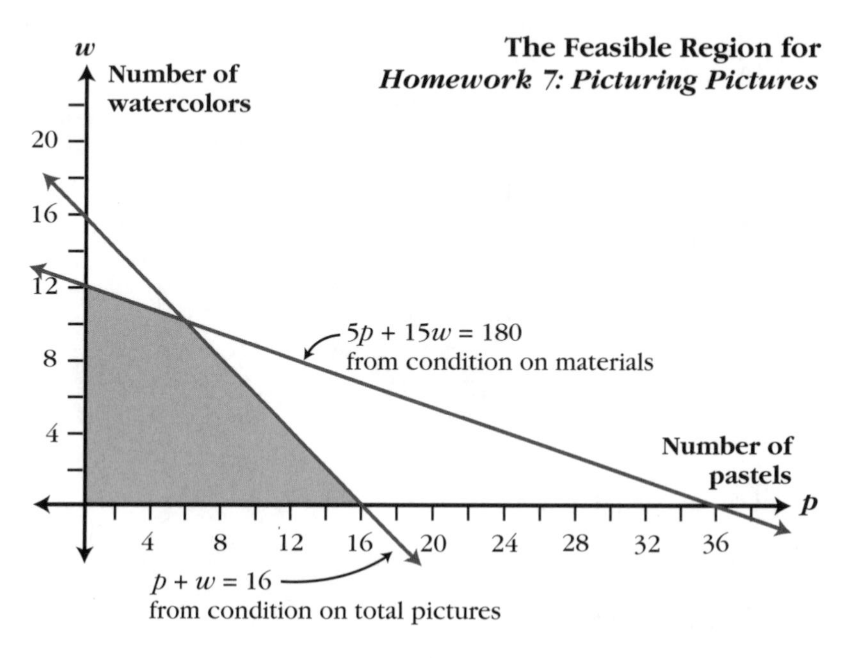

"What does the graph tell you?"

Ask students to articulate what this graph means. They should recognize that each point in the feasible region (or at least each whole-number point) represents a possible choice Hassan can make about how many pictures of each type to make.

Comment: The issue of limiting the variables to whole numbers may come up again here, as it did for the cookie problem. That is, students may point out that they shouldn't be shading the entire region bounded by the lines, but only marking the points with whole-number coordinates. If the issue is raised, you can review the ideas from the earlier discussion of this issue (see the subsection "Whole numbers only?" on Day 6).

- ***Questions 3 and 4: The profit function***

You can have several students each give the profit for one of the points they used in Question 3. You might put this information in an In-Out table or other type of chart as it is presented. Then ask for the profit function in terms of the variables p and w.

Note: Some students may have read the amounts \$40 and \$100 as selling prices rather than profits, because that was the way information was presented in the unit problem. It is important that this issue be clarified before students begin work on today's activity, *Profitable Pictures.*

You may want to look at the various points that students suggested and the profits for these points to determine the maximum profit achieved so far. Tell students that they will learn a way to find the maximum profit and to *prove* that they have the maximum.

2. *Profitable Pictures*

Have students work in their groups on the activity *Profitable Pictures.* They will continue their work on this activity tomorrow. During tomorrow's work, each group will prepare a report on this activity. The report should include answers to Questions 1 through 4 but should be focused especially on Question 5.

Note: If groups have trouble finding combinations that yield a specific profit, you might suggest that they consider points both outside the feasible region and in the region.

Homework 8: Curtis and Hassan Make Choices

This assignment is intended to strengthen students' understanding that profit (or cost) lines form a parallel family. No specific discussion is scheduled for this assignment, but you may want to use students' work on these problems in conjunction with tomorrow's discussion of *Profitable Pictures.*

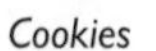

Profitable Pictures

Hassan asked his friend Sharma for advice about what combination of pictures to make. She suggested that he determine a reasonable profit for that month's work and then paint what he needs in order to earn that amount of profit.

Here are the facts you need from *Homework 7: Picturing Pictures.*

- Each pastel requires $5 in materials and earns a profit of $40 for Hassan.
- Each watercolor requires $15 in materials and earns a profit of $100 for Hassan.
- Hassan has $180 to spend on materials.
- Hassan can make at most 16 pictures.

See if you can help Hassan and Sharma. Turn in a written report on the situation. This report should include your work on Questions 1 through 4, but the most important part is your explanation on Question 5.

1. You have already found the feasible region for the problem, which is the set of points that satisfy the constraints. On graph paper, make a copy of this feasible region to use in this problem. Label your axes and show the scales.
2. Suppose Hassan decided $1,000 would be a reasonable profit.
   a. Find three different combinations of watercolors and pastels that would earn Hassan a profit of exactly $1,000.
   b. Mark these three number pairs on your graph from Question 1.

Continued on next page

3. Now suppose Hassan wanted to earn only $500 in profit. Find three different combinations of watercolors and pastels that will earn Hassan a profit of exactly $500. *Using a different-colored pencil,* add those points to your graph.

4. Now suppose that Hassan wanted to earn $600 in profit. Find three different combinations of watercolors and pastels that will earn Hassan a profit of exactly $600. *Using a different-colored pencil,* add those points to your graph.

5. Well, Hassan's mother has convinced him that he should try to earn as much as possible. So Hassan needs to figure out the most profit he can earn within his constraints. He also wants to be able to prove to his mother that it is really the maximum amount.

   a. Find the maximum possible profit that Hassan can earn and the combination of pictures he should make to earn that profit.

   b. Write an explanation that would convince Hassan's mother that your answer is correct.

Homework 8 Curtis and Hassan Make Choices

1. Curtis goes into the pet store to buy a substantial supply of food for his pet. He sees that Food A cost $2 per pound and that Food B cost $3 per pound. Because he intends to vary his pet's diet from day to day anyway, he isn't especially concerned about how much of each type of food he buys.

   a. Suppose that Curtis has $30 to spend. Come up with several combinations of the two foods that he might buy, and plot them on an appropriately labeled graph.

   b. Come up with some combinations that Curtis might buy if he were spending $50, and plot them on the same set of axes used in Question 1a.

   c. What do you notice about your answers to Questions 1a and 1b?

2. Hassan has a feeling there's going to be a big demand for his work. He is considering changing his prices so that he earns a profit of $50 on each pastel and $175 on each watercolor.

   a. Based on these new profits for each type of picture, find some combinations of watercolors and pastels so that Hassan's total profit would be $700, and plot them on a graph. (*Note:* The combinations you give here don't have to fit Hassan's usual constraints.)

   b. Now do the same for a total profit of $1,750, using the same set of axes.

   c. What do you notice about your answers to Questions 2a and 2b?

DAY 9 Continuing to Profit from Pictures

Students use a family of parallel lines to locate the combination that maximizes profit.

Mathematical Topics

- Seeing that setting a linear expression equal to different constants gives a family of parallel lines
- Maximizing a linear function on a polygonal region
- Relating *point of intersection* to *common solution*

Outline of the Day

In Class

1. Select presenters for tomorrow's discussion of *POW 11: A Hat of a Different Color*
2. Provide additional time as needed for students to work on *Profitable Pictures* (from Day 8)
3. Discuss *Profitable Pictures*
   - See that combinations yielding a given profit form a straight line, and introduce the term **profit line**
   - Emphasize that different profits give parallel lines, and discuss why
   - See that profit is maximized at the most extreme point where a member of the profit line family intersects the feasible region
4. Discuss ways to find the point of intersection of the graphs of two linear equations

At Home

Homework 9: Finding Linear Graphs

Special Materials Needed

- A transparency of the feasible region for *Homework 7: Picturing Pictures*

Note: We suggest that you omit having a separate discussion for *Homework 8: Curtis and Hassan Make Choices,* although you might look at this homework briefly as part of the discussion of *Profitable Pictures* (in connection with parallel profit lines).

Discuss With Your Colleagues

Why Not Just Go for the Corners?

Moving the profit lines to find the maximum or minimum takes a lot of time and makes the work tedious for students. Why not just tell them to look at the corners?

1. POW Presentation Preparation

Presentations of *POW 11: A Hat of a Different Color* are scheduled for tomorrow. Choose three students to make POW presentations, and give them overhead transparencies and pens to take home to use in their preparations. *Note:* If students worked on the POW in groups, then choose three groups rather than three individuals.

2. Continuation of *Profitable Pictures*

"What do you notice about combinations that produce a given profit?"

"What happens as the profit increases?"

Have students continue work on the activity from yesterday. As they work on Question 5, you may have to ask some groups leading questions to help them formulate an explanation. For example, you might ask:

- What do you notice about combinations that produce a given profit?
- What happens as the profit increases?

At some point, you might remind groups that they will be preparing written reports for this activity.

When most groups seem to have gotten as far with Question 5 as they are going to get, begin the discussion.

3. Discussion of *Profitable Pictures*

You can start the discussion by putting the transparency of yesterday's graph on the overhead projector and having different students each mark a point from Question 2—a way in which Hassan can earn exactly $1,000. The only whole-number points that give this profit are (0, 10), (5, 8), and (10, 6) (see the next subsection, "The points for each profit are collinear").

Then have students from different groups give the points they used for Question 3, and mark them on the transparency. Be sure to use a different color for this so the points are distinguishable. The only whole-number points in the feasible region that give a profit of exactly $500 are (0, 5), (5, 3), and (10, 1).

Finally, look at Question 4. There are *four* whole-number points in the feasible region that yield a profit of \$600: (0, 6), (5, 4), (10, 2), and (15, 0).

- *The points for each profit are collinear*

"What do you notice about the different groups of points?"

Either as the class does each problem or after all three sets of points are plotted (in different colors for each profit), ask students what they notice about the different groups of points. It is essential for them to recognize that the points for each profit lie on a straight line.

Introduce the term **profit line** for the set of points with a given profit, and have students draw in the complete lines, connecting the individual points they found in Questions 2 through 4. You can extend these profit lines to include points outside the feasible region. The line will also include points whose coordinates are not whole numbers, even though Hassan can't make fractions of pictures.

You may also want to have students check points on each line, including some that are outside the feasible region, to see that they would give the right profit, even if Hassan can't use them. In other words, have students estimate the coordinates of points on the line and see what profit each gives. These profits should be approximately the same as the profit for the points they plotted originally. (If students got the coordinates exactly, the profits would also match exactly.)

As Kara Cannon works on her graph, Leah McLaughlin shares one of her own insights about the graph.

"Why do the points for a given profit lie on a straight line?"

Ask students why the points for a given profit lie on a straight line. If necessary, ask what condition the coordinates must satisfy for a point to give a profit of \$1,000.

Yesterday students identified the profit function as $40p + 100w$. Now they should be able to see that for a combination of pictures to yield a profit of \$1,000, the coordinates must satisfy the equation $40p + 100w = 1000$. That is, they should recognize the set of number pairs that give a profit of \$1,000 as being the same as the set of solutions to the equation $40p + 100w = 1000$, and the points corresponding to these pairs as the graph of the equation.

Once they see the set of points this way, they should be able to identify the equation as being *linear*, and hence explain why the points lie on a straight line.

Note: The diagram should now look something like this:

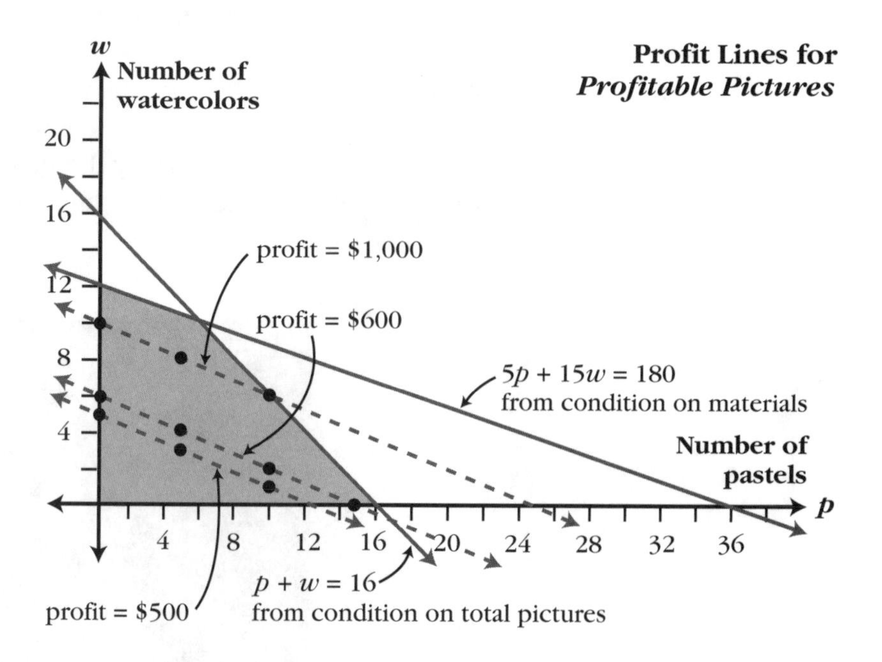

- *Profit lines are parallel*

"What do you notice about the set of profit lines?"

A crucial element of this analysis is that the profit lines are all parallel. If it hasn't yet been mentioned, ask students what they notice about these lines. Also, if they haven't already done so, have students draw in the actual lines and label each with the profit amount it represents. This should make it clearer that these are *parallel lines* and that as the profit increases, the line "slides" upward to the right.

"What does it mean for lines to be parallel?"

As an introduction to a discussion of *why* the lines are parallel, ask students what it means for lines to be parallel. Bring out the key idea that parallel lines have no points in common.

You can then use the algebraic representations of the different lines to reinforce the visual evidence that the lines are parallel. Thus, students should have already seen that the three profit lines have these equations:

$40p + 100w = 1000$ for a profit of $1,000

$40p + 100w = 500$ for a profit of $500

$40p + 100w = 600$ for a profit of $600

Students may reason in various ways to see that these lines are parallel. One approach is to observe that the equations $40p + 100w = 1000$ and $40p + 100w = 500$ cannot have any solutions in common, because the expression $40p + 100w$ cannot be equal to both 1000 and 500. Because the equations have no common solutions, the graphs have no common points, which means the lines are parallel.

Students may also use the similarity among the equations to formulate an explanation, building on their work on the activity *Get It Straight* in *Solve It!* Whatever explanations they give, students should grasp that the lines are genuinely parallel and don't simply *seem* to be parallel on the graph.

Note: The term *slope* will be introduced informally later today (see the subsection "Parallelism and slope"), but the formal definition is not given until the Year 3 unit *Small World, Isn't It?* In that unit, slope is discussed in the context of a general treatment of rates of change. Therefore, you should not initiate that approach here in explaining why the profit lines are parallel.

At this point, you may want to discuss briefly *Homework 8: Curtis and Hassan Make Choices,* to confirm that students got parallel cost or profit lines on those problems. You can have students express each cost or profit line as an equation so that they see the similarity in the algebraic forms.

• *Maximizing profit*

"What is Hassan's maximum possible profit? How can you be sure?"

Now ask several groups to state what they think is the maximum overall profit Hassan can earn and to justify their answer. Roughly, the reasoning should begin with these ideas:

- The points that give a specific profit satisfy an equation that has a form similar to those listed above:

$$40p + 100w = \text{profit}$$

- For any particular profit, these points will lie on a line that is parallel to the three lines students have from Questions 2 through 4.
- Within this family of parallel lines, profit goes up as the line chosen is farther up and to the right.

From here, students should see intuitively that they want to "slide" one of these parallel lines up and to the right until it is about to leave the feasible region. If the sketch is done carefully, they will see that among those lines in the family that actually intersect the feasible region, the most "extreme" line is the one through the point where the two lines $p + w = 16$ and $5p + 15w = 180$ intersect (see the subsection, "Which is the most extreme point?"). Therefore, the point where these two lines intersect represents the maximum profit. Because this reasoning is so visual, most students should be able to understand it, even if they didn't discover it on their own.

"How can you confirm that you have the right coordinates for the intersection point?"

Probably, at least some students will have found the coordinates of this point, (6, 10), either by guess-and-check or by other means. (If not, you can have them do so now, perhaps by estimation from the graph.) No matter how the coordinates are found, you should have students confirm that this point fits both equations.

You can have students find Hassan's profit for this combination of pictures, perhaps comparing it to any previous values that they thought were optimal.

The section "Points of Intersection" later today presents some more ways to handle the task of finding the point of intersection.

• *Which is the most extreme point?*

Because the profit lines are almost parallel to the line $5p + 15w = 180$, it may not be clear from students' graphs where the family of parallel lines leaves the feasible region.

"If it's not clear from the graph, how can you decide which point maximizes profit?"

If this uncertainty is raised, you can ask students what other points seem likely. They should see from the graph that the point (0, 12) is also a reasonable candidate. Ask students how they can be sure which is the right point if they don't trust their graphs.

They should see that they can simply compute the profit at the two points and compare. They will see that the point (6, 10) (representing 6 pastels and 10 watercolors) gives a profit of $1,240 while the point (0, 12) (representing just 12 watercolors) gives a profit of only $1,200.

Note: Some of the upcoming class activities and homework assignments give students an opportunity to look at the issue of how a change in the parameters of a problem can affect the solution.

• *The case of profit lines parallel to a constraint line*

Students may wonder what to do when the family of parallel lines is *parallel* to a side of the feasible region. In that case, all points on that side will give the same profit. You can have an interesting discussion about how one would make a decision in that case.

For example, Hassan might decide which point along that side to use on the basis of what he likes to paint, because the profit is the same. In the unit problem, if one side of the feasible region were parallel to the family of parallel lines, the Woos might decide to choose the point along that side that maximizes the number of plain cookies, because they think that plain cookies are healthier.

Optional: You may want to ask students to look for a pair of profits for each type of picture that would create profit lines parallel to a constraint line. Question 2 of *Homework 11: Changing What You Eat* provides an example of this.

- ***For teachers: What are the goals of this unit?***

Students may soon realize that the feasible region is always a polygon and that the place where the family of parallel lines leaves the region will always be a vertex (or an entire side) of that polygon. But it is *not* a goal of this unit that students learn the principle that the maximum is always at a vertex. Rather, the goal is for students to deepen their understanding of the relationship between equations or inequalities and their graphs and for them to reason and solve problems using graphs.

In particular, students should learn about parallel lines, should reason geometrically, and should use various ways, both algebraic and graphical, to find the common solution to a pair of linear equations. Therefore, when they are solving a particular problem, they should draw on the "family of parallel lines" reasoning, which explains *why* the maximum (or minimum) occurs at a vertex. Simply finding all the vertices and comparing profit (or whatever is being maximized or minimized) should not be considered as sufficient explanation for why a particular point gives the maximum.

- *Parallelism and slope*

In discussing the family of parallel lines, you can describe the lines as "having the same slope" and tell students that **slope** is a mathematical term related to "the amount of slant" a line has. Thus, because parallel lines have "the same amount of slant" we say that they "have the same slope." The actual formula by which we measure "amount of slant" is really not needed in this context.

As noted earlier, slope will be dealt with in detail in the Year 3 unit *Small World, Isn't It?*

4. Points of Intersection

"What is your favorite way for finding the solution to a pair of linear equations?"

Students saw in *Profitable Pictures* that the point they wanted was the place where the two lines $p + w = 16$ and $5p + 15w = 180$ meet. To answer Hassan's question, they needed to find the coordinates of this point of intersection. Ask students to share the different methods they have used to do this.

In this problem, the equations are simple enough that guess-and-check probably sufficed for students to find that values of 6 and 10 for p and w, respectively, give the point that the two lines have in common. You can encourage students to use guess-and-check as a reasonable first approach, but point out that in other situations, guess-and-check might not suffice (especially if the solution involves fractions).

Another good approach to finding the coordinates is estimation from the graph. Because the coordinates were actually whole numbers in Hassan's problem, this method would give the exact value.

If students find the numbers by guess-and-check with the equations, you can urge them to check that their answer looks like a good approximation to the coordinates of the point of intersection on the graph. On the other hand, if they find the numbers by estimating the coordinates from the graph, they should check that the numbers do actually satisfy both equations.

Whatever method students use to find this point, you can bring out that the values they find for p and w have two distinct but closely related properties:

- These numbers are the p- and w-coordinates of the point where the two lines intersect.
- These numbers are the values for p and w that satisfy both of the equations.

In other words, use this opportunity to make sure that students understand the connection between the concepts of equation and graph.

Homework 9: Finding Linear Graphs

This homework and tomorrow's discussion should aid students who are having trouble graphing equations.

Homework 9 Finding Linear Graphs

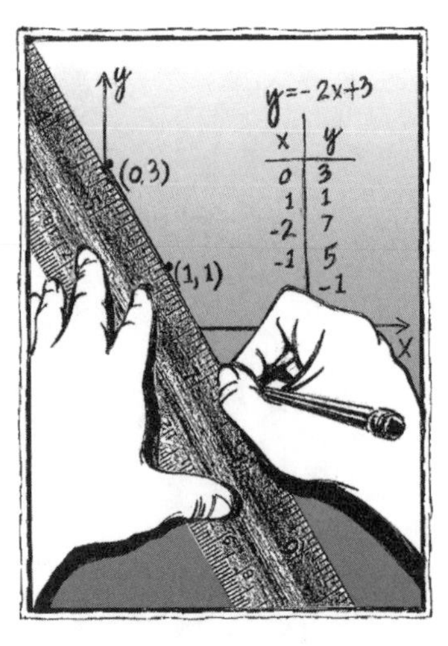

Throughout this unit, you are using the graphs of linear equations and inequalities to understand problems. The purpose of this assignment is to have you look at the techniques you use to graph linear equations, and perhaps to find some shortcuts. In class tomorrow, you can share what you have found with others.

1. One approach to graphing is to make a table of number pairs that fit the equation, graph them, and then connect the points with a straight line.
   a. Create a table of at least five number pairs that satisfy the equation $3x + y = 9$.
   b. Plot the number pairs from your table and connect them with a straight line.

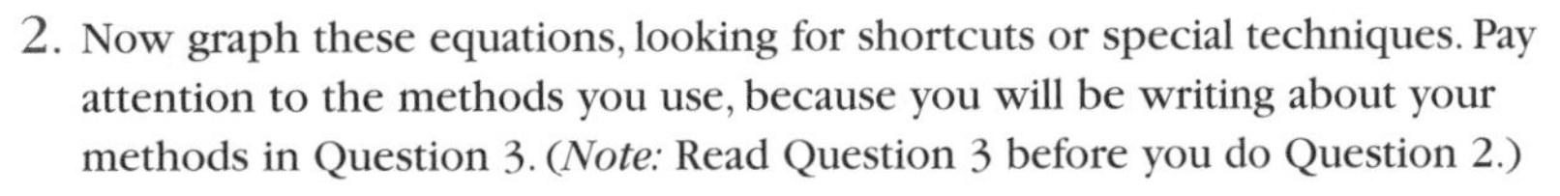

2. Now graph these equations, looking for shortcuts or special techniques. Pay attention to the methods you use, because you will be writing about your methods in Question 3. (*Note:* Read Question 3 before you do Question 2.)
   a. $y = x + 4$
   b. $x + y = 6$
   c. $2x = 3y$
   d. $2x + 3y = 12$
   e. $5y = 6x - 30$

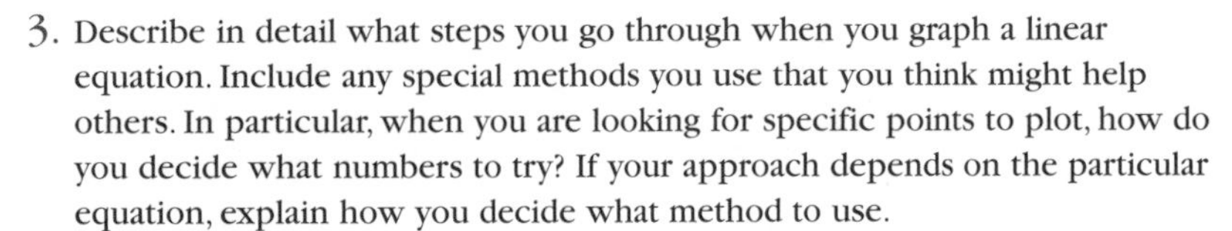

3. Describe in detail what steps you go through when you graph a linear equation. Include any special methods you use that you think might help others. In particular, when you are looking for specific points to plot, how do you decide what numbers to try? If your approach depends on the particular equation, explain how you decide what method to use.

DAY 10

POW 11 Presentations

After POW presentations, students share ideas about graphing linear equations.

Mathematical Topics

- Using logical reasoning
- Summarizing methods for graphing linear equations
- Maximizing various functions over a feasible region

Outline of the Day

In Class

1. Presentations of *POW 11: A Hat of a Different Color*
2. Discuss *Homework 9: Finding Linear Graphs*
   - Have students share strategies for graphing linear equations
3. *Hassan's a Hit!*
   - Students see that changing the profit function can change the point that gives the maximum profit
   - The activity will be discussed on Day 11

At Home

Homework 10: You Are What You Eat

POW 12: Kick It! (due Day 18)

1. Presentations of *POW 11: A Hat of a Different Color*

Have the three students (or three groups) make their presentations. Be aware that you may have very different levels of understanding of this problem.

The key is recognizing that certain possibilities are eliminated by the fact that neither Arturo nor Belicia can figure out the color of their hat.

For example, had Arturo seen two red hats, he would have known that his was blue (because there are only two red hats). That tells both Belicia and Carletta that at least one of them has a blue hat. Similar reasoning from the fact that Belicia doesn't know what color hat she has gives Carletta the answer.

2. Discussion of *Homework 9: Finding Linear Graphs*

You can focus the discussion on Question 3 by asking students to share ideas in their groups on how they approach graphing linear equations. Then ask each diamond card student to share an idea with the class. Keep rotating among the groups until no one has any additional ideas to offer.

Be sure that all students grasp the general idea that they get the graph by finding some points that satisfy the equation and then connecting those points with a straight line. Some students may realize that two points will suffice. Others may be aware that it's often good to get three points, as a check in case one of them is incorrect.

"How do you come up with specific points to plot?"

If no one brings up the idea that often the easiest points to find are the two intercepts, refer to the part of Question 2 that asked students how they found specific points to plot.

It is not necessary that students learn any particular method, as long as they know at least *one* method for getting the graph. But they will be graphing lots of linear equations during the unit (and in other settings as well), so it's only fair to give them a chance to share clever ways to do this.

3. *Hassan's a Hit!*

Students often believe that once they have graphed the feasible region, they can maximize profit by finding the point in the region that is farthest from the origin. The main purpose of this problem is to deter them from being so hasty and to convince them that they really need to draw the profit lines to find the maximum profit.

Ask students to work on the assignment in groups. It will be discussed tomorrow. Give overhead transparencies to a couple of groups that finish early, so they can make presentations tomorrow.

Homework 10: You Are What You Eat

Students will probably approach this problem using the techniques of the past few days, although they can use a more intuitive approach.

POW 12: Kick It!

This problem gives students a chance to explore ways of combining numbers. They should come up with some conclusions related to common divisors. You may want to point out that only whole-number scores are to be considered.

This POW provides considerable opportunity for exploration and generalization. Although students should be looking for patterns among their results, they are also expected to prove some of them. Be sure they are aware, for example, that Question 1 involves *proving* that all scores above a certain value can be achieved using field goals and touchdowns. This is a good problem with which to make the distinction between showing lots of examples and giving a general proof.

You may want to assure students that they do not need to know anything about football or its scoring system in order to work this problem. They only need to know that in this problem, there are two ways to score points: *field goals* and *touchdowns*.

If you think some additional information will make students more comfortable, you can add that field goals involve kicking the ball while touchdowns involve running with it or throwing it. But be careful not to overwhelm students who are unfamiliar with football with more information than they need. For example, you don't need to get into the fact that in standard football, there are other ways to score points.

This POW is scheduled for discussion on Days 18 and 19. You may want to check in a day or two to see whether students need some in-class time to work on the POW.

Hassan's a Hit!

Hassan's pictures are indeed a big hit, especially the watercolors. Based on his success, he is raising his prices as he planned in *Homework 8: Curtis and Hassan Make Choices.* That is, he will now earn a profit of \$50 on each pastel and \$175 on each watercolor.

Assume that Hassan still has the same constraints. That is, he still has only \$180 to spend on materials and can make at most 16 pictures. He had already figured out, with the old prices, how many of each type of picture he should make to maximize his overall profit.

If Hassan wants to maximize his overall profit with the new prices, should he now change the number of pictures he makes of each type? Explain your answer.

Homework 10 You Are What You Eat

The Hernandez twins do not like breakfast. Given a choice, they would rather skip breakfast and concentrate on lunch.

When pressed, the only things they will eat for breakfast are Sugar Glops and Sweetums cereals. (The twins are allergic to milk, so they eat their cereal dry.)

Mr. Hernandez, on the other hand, thinks his children should eat breakfast every single morning. He also believes that their breakfast should be nutritious. Specifically, he would like them each to get at least 5 grams of protein and not more than 50 grams of carbohydrate each morning.

According to the Sugar Glops package, each ounce of that cereal has 2 grams of protein and 15 grams of carbohydrate. According to the Sweetums box, each ounce of that cereal contains 1 gram of protein and 10 grams of carbohydrate.

So what's the least amount of cereal each twin can eat while satisfying their father's requirements? (Mr. Hernandez wants a proof that his criteria are met, and the twins want a proof that there's no way they can eat less.)

POW 12 *Kick It!*

The Free Thinkers Football League simply has to do things differently. The folks in this league aren't about to score their football games the way everyone else does. So they have thought up this scoring system:

- Each field goal counts for 5 points.
- Each touchdown counts for 3 points.

The only way to score points in their league is with field goals or touchdowns or some combination of them.

Continued on next page

One of the Free Thinkers has noticed that not every score is possible in their league. For example, a score of 1 point isn't possible, and neither is 2 or 4. But she thinks that beyond a certain number, all scores are possible. In fact, she thinks she knows the highest score that is impossible to make.

1. Figure out what that highest impossible score is for the Free Thinkers Football League. Then explain why you are sure that all higher scores are possible.
2. Make up some other scoring systems (using whole numbers) and see whether there are scores that are impossible to make. Is there always a highest impossible score? If you think so, explain why. If you think there aren't always highest impossible ones, find a rule for when there are and when there are not.
3. In the situations for which there is a highest impossible score, see if you can find any patterns or rules to use to figure out what the highest impossible score is. You may find patterns that apply in some special cases.

Write-up

1. *Problem Statement*
2. *Process:* Include a description of any scoring systems you examined other than the one given in the problem.
3. *Conclusions:*
   a. State what you decided is the highest impossible score for the Free Thinkers' scoring system. Prove both that this score is impossible and that all higher scores are possible.
   b. Describe any results you got for other systems. Include any general ideas or patterns you found that apply to all scoring systems, and prove that they apply in general.
4. *Evaluation*
5. *Self-assessment*

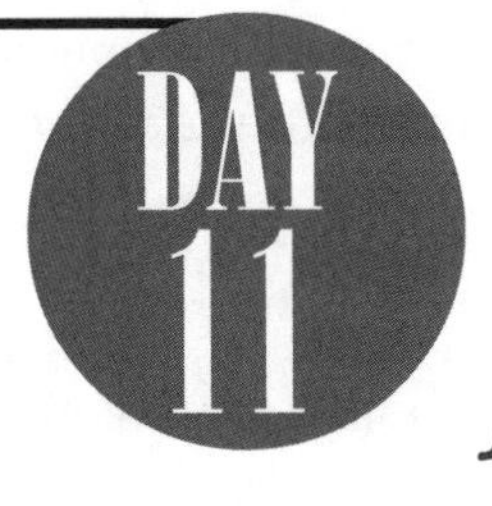

Eating a Minimal Breakfast

Students apply their new methods to minimizing a linear expression.

Mathematical Topics

- Minimizing a variable subject to simple constraints
- Seeing that changing the profit function can change the point that gives the maximum profit

Outline of the Day

In Class

1. Discuss *Homework 10: You Are What You Eat*
   - Students set up inequalities, find the feasible region, and identify the solution
2. Discuss *Hassan's a Hit!* (from Day 10)
   - Students see that the point of maximum profit depends on the profit function and not simply on the feasible region

At Home

Homework 11: Changing What You Eat

1. Discussion of *Homework 10: You Are What You Eat*

"What combination of cereals with the least total amount will satisfy Mr. Hernandez?"

Ask each group to decide on the combination with the least amount of cereal that will satisfy Mr. Hernandez. Then choose one group to present its solution.

Students will probably set this up like their previous problems. If they use x for the number of ounces of Sugar Glops and y for the number of ounces of Sweetums, then the constraints are:

$2x + y \geq 5$ (to guarantee enough protein)

$15x + 10y \leq 50$ (to avoid too much carbohydrate)

$x \geq 0, y \geq 0$ (because the amounts can't be negative)

Students should also identify what quantity they are minimizing—the twins' goal is to minimize $x + y$.

The shaded area in the graph here shows the feasible region for this problem. The dashed line is a sample "consumption line" showing combinations for which the twins eat a total of 1 ounce of cereal.

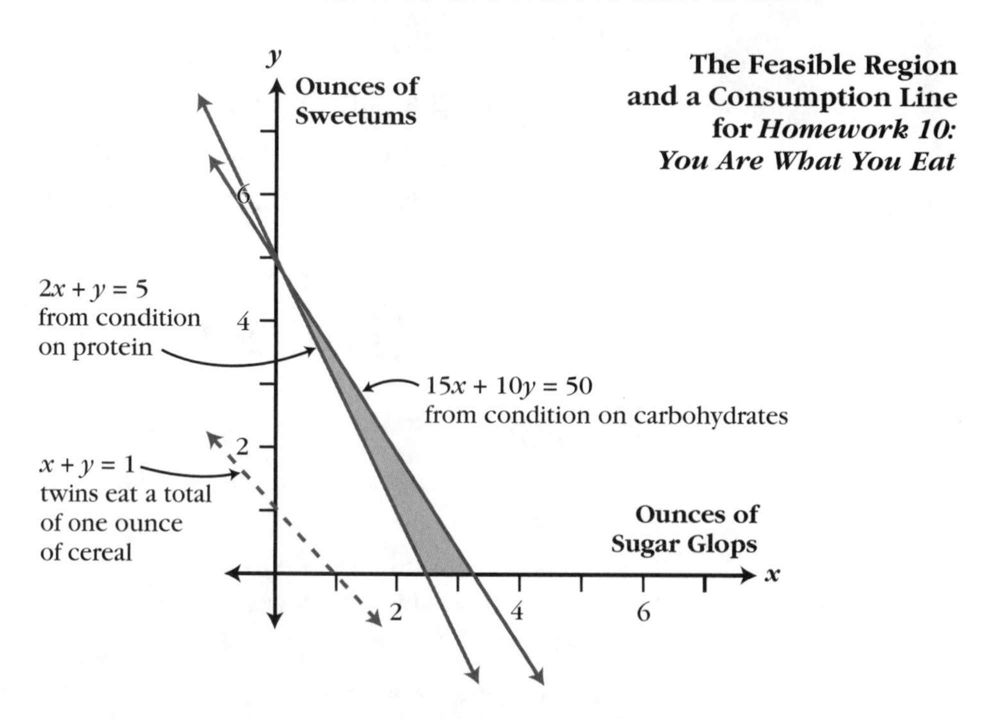

The "lowest" consumption line that intersects the feasible region is the one that goes through the point $(2\frac{1}{2}, 0)$, which is where the line $2x + y = 5$ meets the x-axis. Therefore, $2\frac{1}{2}$ ounces of Sugar Glops (and no Sweetums) would be the best solution from the twins' point of view.

"Why doesn't the carbohydrate constraint play a role in the final solution?"

You can point out that the carbohydrate constraint does not play a role in the final solution to the problem and ask students why this is the case. They should say something to the effect that the limit of 50 grams of carbohydrate is high enough that the twins get their requirement of protein without getting close to the carbohydrate limit. Because they want to eat as little as possible anyway, the carbohydrate condition is not a problem.

It may be worthwhile to have students look at how the graph reflects the fact that the carbohydrate condition is immaterial. The pair of cereal quantities must be at or above the "protein line" in order for the twins to get enough protein. Because the point on this line that uses the least cereal is $(2\frac{1}{2}, 0)$, all that matters as far as carbohydrates are concerned is that this point be at or below the "carbohydrate line," which it is.

Students may come up with explanations for this solution without making a graph or formally identifying $x + y$ as the quantity to be minimized. For example, they may point out that the twins get more protein per ounce from Sugar Glops than from Sweetums, so there is no reason to eat any Sweetums. If they present this reasoning, they need to check that $2\frac{1}{2}$ ounces of Sugar Glops does not provide too much carbohydrate.

"Did anything like this happen in the unit problem?"

Ask students if they recall seeing anything similar happen in the unit problem. If necessary, you can refer them to the graph of the feasible region (posted on Day 7) to see that the line $P + I = 140$, from the oven-space constraint, was completely outside the region. (See the subsection "An extraneous constraint" on Day 7.)

2. Discussion of *Hassan's a Hit!*

Ask the groups to which you gave overhead transparencies yesterday to make a presentation. The feasible region is the same as before, but now the expression for profit is $50p + 175w$.

As in the original problem, students can compare profits at specific points to confirm their graphical analysis. In this case, the profit at (6, 10) is \$2,050, while the profit at (0, 12) is \$2,100.

The diagram here shows the feasible region together with three of these new profit lines.

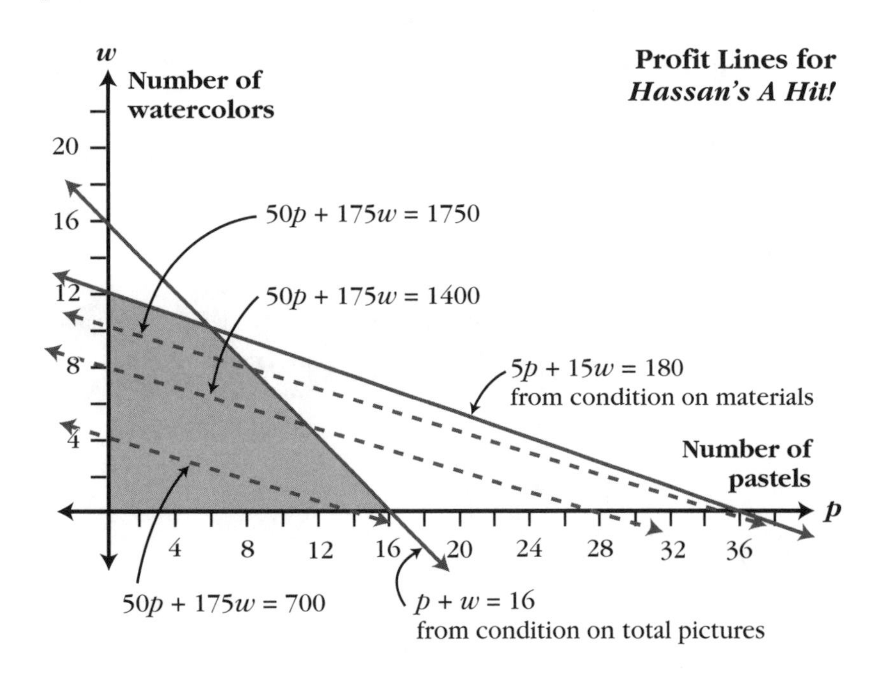

By drawing these profit lines on their old feasible region, students will probably see that these lines seem less steep than the graph of the equation $5p + 15w = 180$. Because of this, as the profit increases, the last profit line to touch the feasible region meets it at (0, 12), and not at (6, 10) as in the original problem.

"But shouldn't the point giving the maximum profit be the point in the region farthest from the origin?"

It is important that all students see that the slope of the parallel profit lines can make different points in the feasible region be a maximum. Thus, drawing the feasible region does not by itself give the answer. To drive this point home, ask, "But shouldn't the point giving the maximum profit be the point in the region farthest from the origin?"

- *Optional: Changing profits to get a different maximum*

You may want to have students experiment with the profit amounts for each type of picture to see how changing these profit values affects which combination of pictures gives the maximum total profit. In the original problem, the maximum total profit was at (6, 10), and in *Hassan's a Hit!* the maximum profit was at (0, 12). You might have students try to find profit values for which the maximum total profit is at (16, 0).

As a real challenge, ask students if they can find a pair of profit values for the two types of pictures so that the point (8, 8) gives the maximum total profit. (There is no way for this combination of pictures to give a *unique* maximum. However, if the profit lines are parallel to the line $p + w = 16$, then no combination will give a *greater* profit than this. To get the profit lines parallel to $p + w = 16$, the two types of pictures simply need to give the same profit.)

Note: Tonight's homework assignment will give students another opportunity, using a different situation, to look at the issue of how a change in the parameters of a problem can affect the solution.

Homework 11: Changing What You Eat

Tonight's homework asks students to examine the way the specific numbers affect the solution to the cereal problem.

Homework 11 Changing What You Eat

In *Homework 10: You Are What You Eat,* the twins' solution was simply to eat Sugar Glops. That way, they could get their protein by eating only $2\frac{1}{2}$ ounces of cereal and still not get too many grams of carbohydrate. But what if the cereals had been a little different from the way they were in that problem, or if Mr. Hernandez had been stricter about the twins' carbohydrate intake, or . . . ?

Here are some specific variations for you to work on.

1. Suppose that Sugar Glops is the same as in the original problem (with 2 grams of protein and 15 grams of carbohydrate per ounce), but now Sweetums also has 2 grams of protein per ounce (and still only 10 grams of carbohydrate per ounce). Also suppose that Mr. Hernandez still has a 50-gram limit on carbohydrate and wants each of the twins to get at least 5 grams of protein.

   How much of each cereal should the twins eat if they want to eat as little cereal as possible?

2. Now suppose that Sugar Glops has 3 grams of protein and 20 grams of carbohydrate per ounce, while Sweetums is the same as in Question 1 (2 grams of protein and 10 grams of carbohydrate per ounce). Also suppose that Mr. Hernandez has now decided that the twins can't eat more than 30 grams of carbohydrate (but they still need at least 5 grams of protein).

   What should the twins do?

Changing What You Eat

Students examine variations on the breakfast problem.

Mathematical Topics

- Examining how changing parameters in a problem changes the solution
- Summarizing how to find a maximum or minimum using a feasible region and a family of parallel lines

Outline of the Day

In Class

1. Discuss *Homework 11: Changing What You Eat*
   - Focus on how changing parameters changes the solution
2. Have students summarize how to use the feasible region and a family of parallel lines to find a maximum or minimum
3. (Optional) If time allows, let students create variations to the breakfast problems that fit certain conditions
4. Introduce *Homework 12: Rock 'n' Rap*

At Home

Homework 12: Rock 'n' Rap

1. Discussion of *Homework 11: Changing What You Eat*

You may want to let students spend a few minutes in their groups comparing results and problems, and then have club card students present the various problems.

• *Question 1*

After some specific combinations have been suggested, you can ask: "Will any $2\frac{1}{2}$-ounce combination work?"

In Question 1, students should see that the twins should eat $2\frac{1}{2}$ ounces of cereal but that any $2\frac{1}{2}$-ounce combination will do. That is, any such combination represents the minimum amount of cereal that provides enough protein and not too much carbohydrate.

The diagram here shows the feasible region for this problem. In this case, one of the constraint lines, $2x + 2y = 5$, is parallel to the family of consumption lines. Therefore, any point along this line represents the minimum amount of cereal.

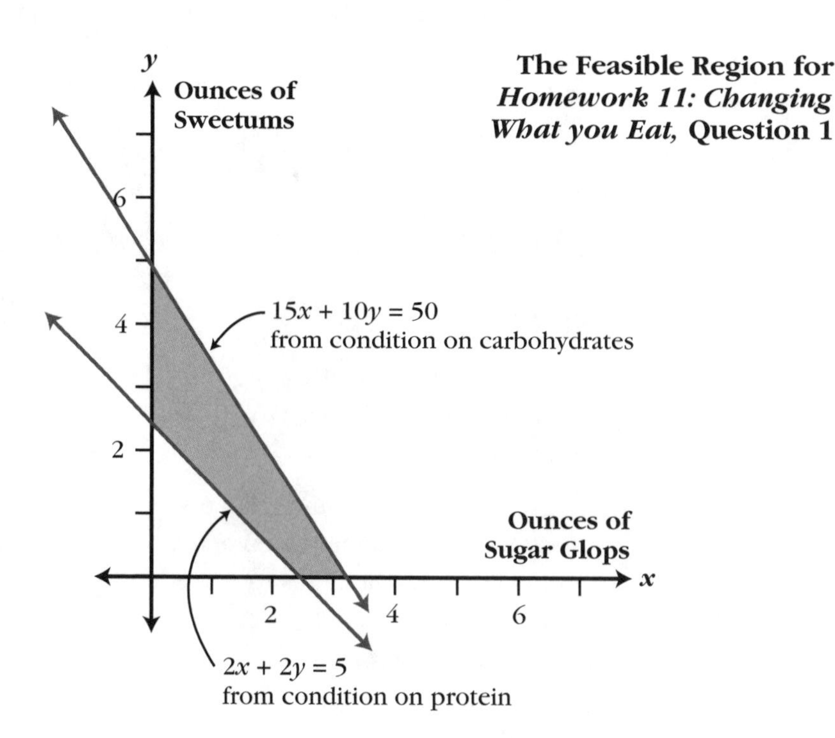

The Feasible Region for *Homework 11: Changing What you Eat,* Question 1

• *Question 2*

"Why can't the twins simply eat $1\frac{2}{3}$ ounces of Sugar Glops?"

In Question 2, however, the twins can't eat only Sugar Glops, even though $1\frac{2}{3}$ ounces of Sugar Glops provides enough protein with the minimum amount of cereal, because $1\frac{2}{3}$ ounces of Sugar Glops has too much carbohydrate. Therefore, they need to include some Sweetums, which has less carbohydrate per ounce. It turns out that exactly 1 ounce of each cereal meets the requirements with the least total cereal, as this diagram of the feasible region shows:

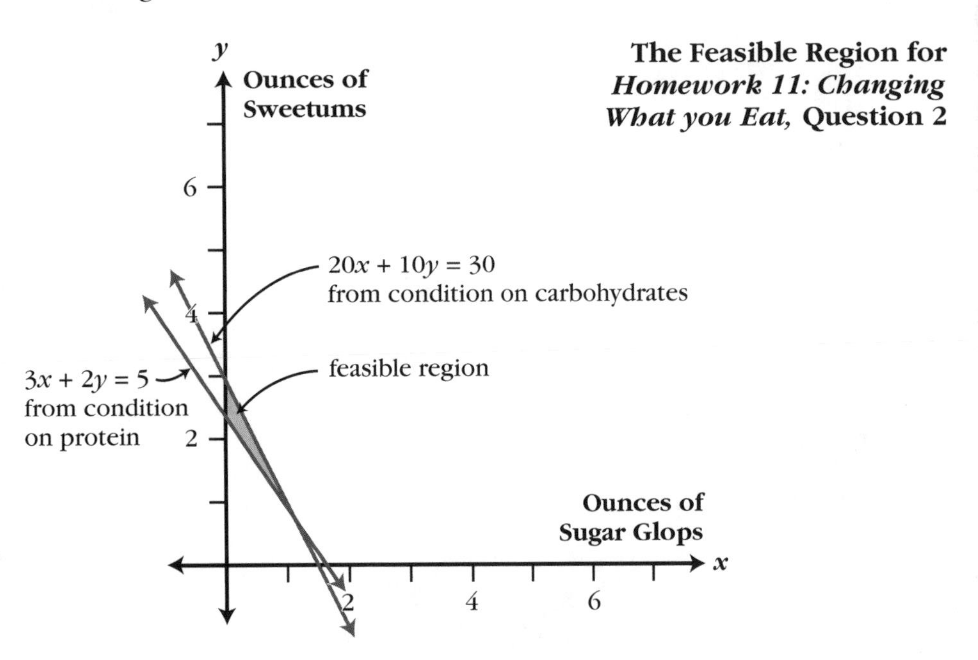

2. Summarizing the Method

Now that students have worked with several feasible regions, it's a good time to ask them to summarize what they've been doing. You might first compile a list of the various situations for which they have used the same basic reasoning, which includes *Profitable Pictures* (Day 8), *Hassan's a Hit!* (Day 10), *Homework 10: You Are What You Eat,* and *Homework 11: Changing What You Eat.*

Ask them to come up with a list showing the reasoning and the steps involved in using a family of parallel lines to find a maximum or minimum. You might have them work in groups for a while and then bring the class together as a whole.

Here is a sample of what to look for. This list is phrased in terms of profit lines, but a list for consumption lines (such as for the breakfast problems) would be similar.

- Graph the feasible region.
- Graph some combinations for a given profit to get a straight line.
- Vary the particular profit to get a family of parallel lines.
- See that as the profit increases, the parallel line shifts up and to the right.

- Look for a line in the family that is farthest "up and to the right" but that still intersects the feasible region.
- Identify the point where this line crosses the region as the desired point.

3. Optional: More Variations on What You Eat

You should leave about 10 minutes for students to get started on tonight's homework. However, as time allows, you can have students work on the supplemental problem *More Cereal Variations,* which poses some further questions about the breakfast situation.

4. Introduction to *Homework 12: Rock 'n' Rap*

Tonight's assignment introduces a new problem situation on which students can apply the concepts and techniques developed so far. This problem is used on Day 14 as a vehicle for exploring the use of the graphing calculator.

Allow about 10 minutes in class for students to read the assignment and ask questions concerning any parts they find unclear. The condition "they will not release more rap music than rock" is often confusing to students, perhaps because it is stated negatively. You may want to suggest that they look at numerical cases to clarify how to express this as an inequality.

Homework 12 Rock 'n' Rap

The Hits on a Shoestring music company is planning its next month's work. The company makes CDs of both rock and rap music.

It costs the company an average of $15,000 to produce a rock CD and an average of $12,000 to produce a rap CD. (The higher cost for rock comes from needing more instrumentalists for rock CDs.) Also, it takes about 18 hours to produce a rock CD and about 25 hours to produce a rap CD.

The company can afford to spend up to $150,000 on production next month. Also, according to its agreement with the employee union, the company will spend at least 175 hours on production.

Hits on a Shoestring earns $20,000 in profit on each rock CD it produces and $30,000 in profit on each rap CD it produces. But the company recently promised its distributor that it would not release more rap music than rock, because the distributor thinks the company is more closely associated with rock in the public mind.

The company needs to decide how many of each type of CD to make. *Note:* It can make a fraction of a CD next month and finish it the month after.

1. Graph the feasible region.
2. a. Find at least three combinations of rock and rap CDs that would give the company a profit of $120,000, and mark these points in one color on your graph. (The combinations do not have be in the feasible region.)

   b. In a different color, mark points on your graph that will earn $240,000 in profits.
3. Find out how many CDs the company should make of each type next month to maximize its profit.
4. Explain how you found an answer to Question 3 and why you think your answer gives the maximum profit.

DAY 13 Rock 'n' Rap

Students look at a new problem about two kinds of CDs.

Mathematical Topics

- Seeing how a problem changes when the parameters are changed
- Working with three constraints in a linear programming problem

Outline of the Day

In Class

1. Discuss *Homework 12: Rock 'n' Rap*
   - Review that the point of intersection of two lines is the same as the common solution to the two equations
2. *A Rock 'n' Rap Variation*
   - Students see that using a different profit function may lead to a different maximum point
3. Discuss *A Rock 'n' Rap Variation*

At Home

Homework 13: Getting on Good Terms

1. Discussion of *Homework 12: Rock 'n' Rap*

Note: Students will not be discussing the unit problem again until Day 19, and the most recent reference to it was on Day 7, so you may want to assure them they will get back to the problem and that they will be learning more about how to solve it and similar problems.

"What are the constraints in the problem?"

You can begin the discussion by asking several spade card students to give constraints for this problem.

You should agree as a class on what variables to use and which variable will go on which axis. We will use x to represent the number of rock CDs produced and y to represent the number of rap CDs produced, with the usual assignment of axes.

Students should come up with constraints that are equivalent to these:

$15{,}000x + 12{,}000y \leq 150{,}000$	(for the amount of money available for production)
$18x + 25y \geq 175$	(for the amount of production time the company owes the union)
$y \leq x$	(as promised to the distributor)
$x \geq 0,\ y \geq 0$	(because the numbers can't be negative)

You might have students simplify the first equation. Either $15x + 12y \leq 150$ or $5x + 4y \leq 50$ is easier to work with than $15{,}000x + 12{,}000y \leq 150{,}000$.

Then you can have another spade card student draw the feasible region on the overhead, showing how the inequalities already given lead to the desired region. The shaded area in the diagram here shows the feasible region for this problem.

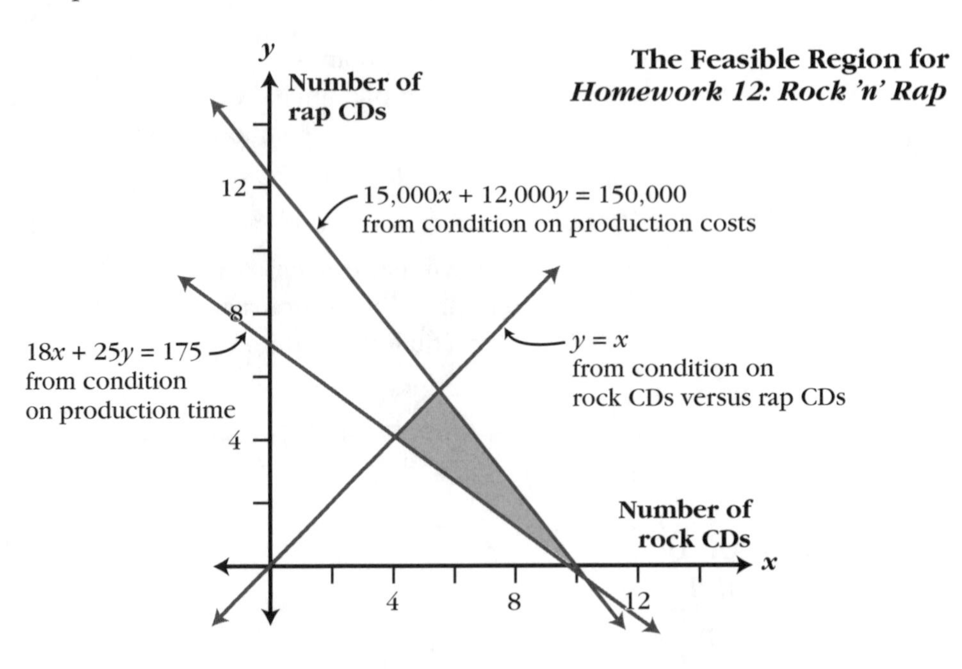

- *Maximizing profit*

"What expression represents profit?"

Once students have drawn the region and discussed it, have the class identify the expression that represents the profit in terms of x and y (namely, $20{,}000x + 30{,}000y$).

Next, ask two students to mark the points for Question 2a on the graph. *Note:* Points (6, 0), (3, 2), and (0, 4) are the only points in the first quadrant with whole-number coordinates that give a profit of $120,000.

Then ask another student to mark the points for Question 2b in a different color on the graph. *Note:* Points (12, 0), (9, 2), (6, 4), (3, 6), and (0, 8) are the only points in the first quadrant with whole-number coordinates that give a profit of $240,000.

"What point maximizes profit?"

Then ask one or more students to explain how to find the point on the graph that will maximize profit. This has two aspects:

- Explaining why the desired point is at the intersection of the two lines $y = x$ and $15{,}000x + 12{,}000y = 150{,}000$
- Finding the coordinates of this point of intersection

Here are some ideas on how students might handle each part.

- ***Identifying the point***

To explain the location of the desired point, students will probably use the "family of parallel lines" reasoning, building on their work in Question 2. The graph here shows the feasible region and the profit lines $20{,}000x + 30{,}000y = 120{,}000$ and $20{,}000x + 30{,}000y = 240{,}000$.

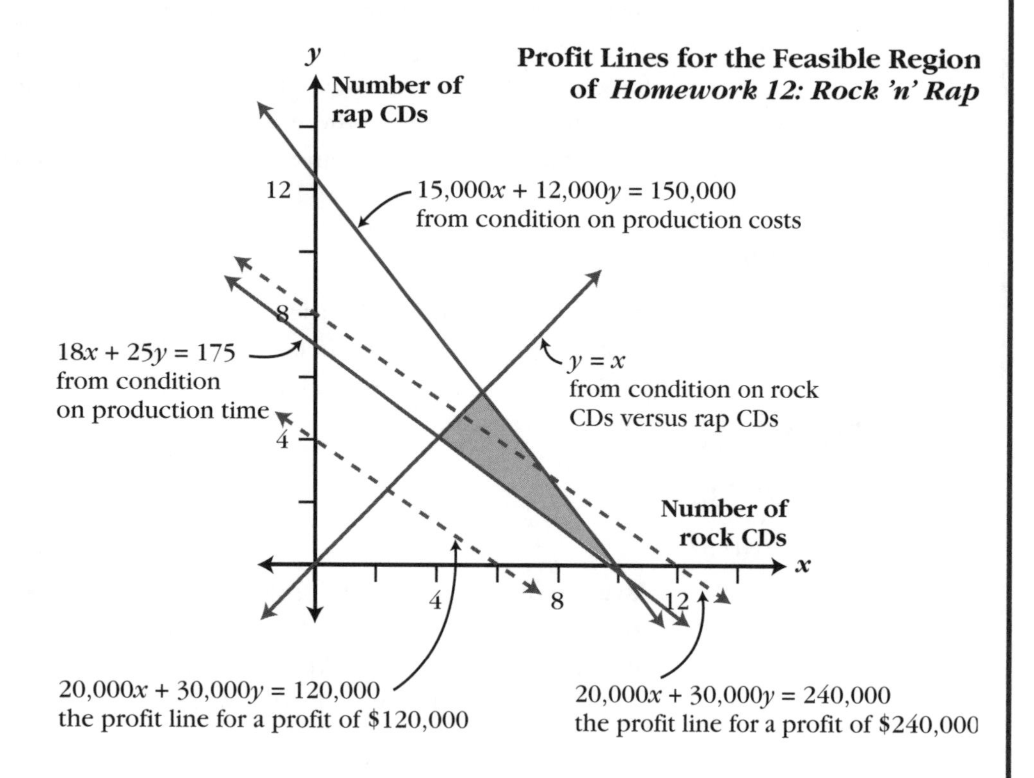

"How does the profit line change as the profit increases?"

Based on a diagram like this, students should see that as the profit increases, the profit line moves up and to the right, and last hits the feasible region at the point where the lines $y = x$ and $15{,}000x + 12{,}000y = 150{,}000$ intersect.

• *Finding the coordinates*

"How do you find the coordinates of this point of intersection?"

Once students have explained where the point of maximum profit is, you can turn to their explanations of how to find the coordinates of this point. Probably, some students will do this graphically (by drawing careful graphs and estimating the coordinates, or by using a graphing calculator), while others will reason it out algebraically. (Because one of the equations is $y = x$, the algebra is particularly simple. Students will look at the general problem of finding coordinates of points of intersection in the activity *Get the Point,* which they begin on Day 15.)

The point of intersection has coordinates $(5\frac{5}{9}, 5\frac{5}{9})$, which students might approximate as (5.6, 5.6). The issue of using decimals to approximate fractional answers is discussed in the subsections "The issue of decimal approximation" on Day 14 and "The decimal approximation issue again" on Day 17.

At the risk of belaboring the issue, get students to articulate the fact that $(5\frac{5}{9}, 5\frac{5}{9})$ represents both of the following:

- The coordinates of the point where the lines meet
- The common solution to the equations $y = x$ and $15{,}000x + 12{,}000y = 150{,}000$

That is, check that students are keeping in mind the connection between an equation and its graph. They should be aware that a point is on a graph if and only if its coordinates satisfy the equation. Thus, in *Homework 12: Rock 'n' Rap,* the point where the lines meet has coordinates that satisfy both equations. This idea can never be mentioned too many times.

2. *A Rock 'n' Rap Variation*

This activity continues the theme of examining how variations in the parameters of a problem affect the solution. As students work on it, choose a group to prepare a presentation.

3. Discussion of *A Rock 'n' Rap Variation*

Have the heart card student from the group you selected make the presentation. The presenter should point out that the feasible region is the same as in *Homework 12: Rock 'n' Rap,* so all that needs to be done is to examine the new family of profit lines.

As the next diagram shows, if the profits are reversed, the profit line has a different slope, and the last point to be "hit" by a profit line is the point where the line $15{,}000x + 12{,}000y = 150{,}000$ meets the x-axis. This point has coordinates (10, 0), so the company should make only rock CDs, and its profit is \$300,000.

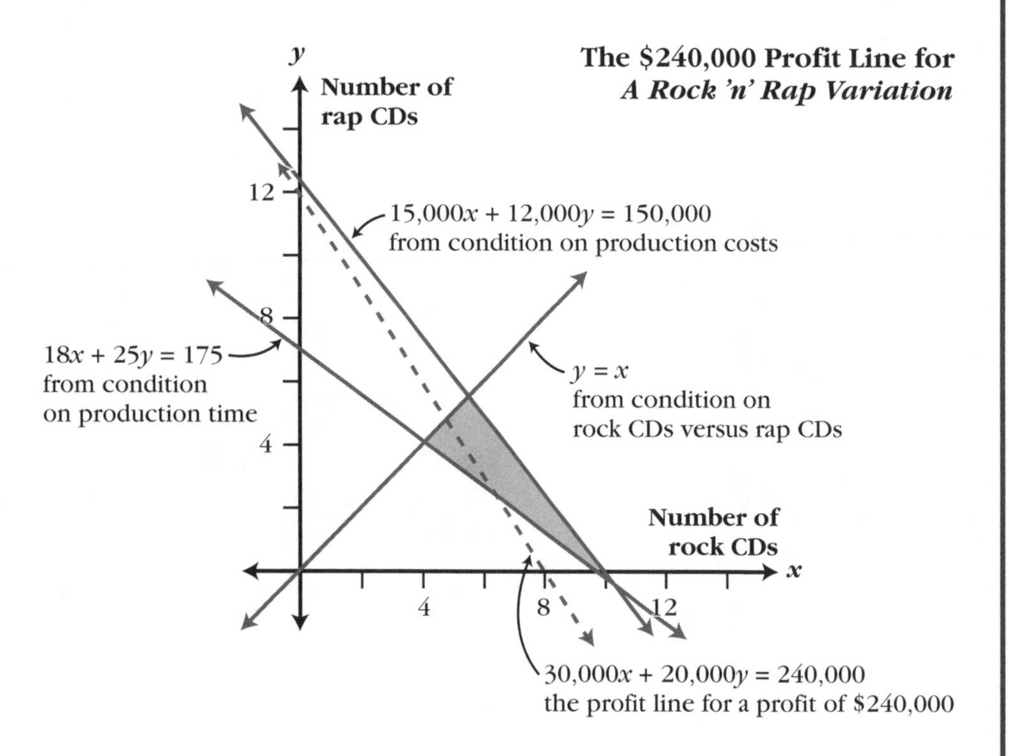

Homework 13: Getting on Good Terms

Tonight's homework has two goals:

- It provides some algebra review.
- It focuses on the process of solving for one variable in terms of another, which students need to know in order to graph certain equations on the graphing calculator.

The discussion tomorrow will also review the concept of equivalent equations, as students see that solutions to their new equations are also solutions to the original equations.

A Rock 'n' Rap Variation

In *Homework 12: Rock 'n' Rap,* you figured out how many rock CDs and how many rap CDs Hits on a Shoestring should produce to maximize its profit.

Suppose the conditions were the same as in that problem except that the profits were reversed. In other words, suppose the company made $30,000 profit on each rock CD and $20,000 profit on each rap CD.

Would this change your advice to the company about how many CDs of each type to produce to maximize its profit? If so, how many of each type should the company make, and what would be the profit? Explain your answer.

Homework 13 Getting on Good Terms

Graphing calculators can make it easier to find feasible regions, but in order to draw the graph of an equation on a graphing calculator, the equation needs to be put into "$y =$" form. That is, you need to write the equation so that one variable is expressed in terms of the other. For example, you might rewrite the equation $y - 5 = 4x$ as $y = 4x + 5$.

For each of the equations here, express the variable y in terms of the variable x.

1. $y - 2x = 7$
2. $7y = 14x - 21$
3. $5x + 3y = 17$
4. $5(x + 3y) = 2x - 3$
5. $4x - 7y = 2y + 3x$
6. $3y + 7 = 20 - (4x - y)$

DAY 14 Calculator Intersections

Students use graphing calculators to draw feasible regions.

Mathematical Topics

- Solving a linear equation for one variable in terms of another
- Using a graphing calculator to draw a feasible region and to estimate the coordinates of the point of intersection of linear graphs

Outline of the Day

In Class

1. Discuss *Homework 13: Getting on Good Terms*
   - Use the concept of equivalent equations to check the work of solving for one variable in terms of another
2. Solve *Homework 12: Rock 'n' Rap* using the graphing calculator
   - Graph the equations that correspond to the inequalities
   - Graph profit lines
   - Solve a variation on the problem

At Home

Homework 14: Going Out for Lunch

1. Discussion of *Homework 13: Getting on Good Terms*

Have students compare answers in their groups, and then have diamond card students from different groups present their methods for solving for y in each equation.

• *Equivalent equations*

"What relationship do the new 'y =' equations have to the original equations?"

"What does 'equivalent equations' mean? How are the graphs of equivalent equations related?"

Either partway through this process or after all the equations have been done, ask the class what relationship the new "$y =$" equations have to the original equations. As needed, remind students of the term *equivalent equations,* and ask for a volunteer to explain what that means in terms of graphs. Bring out that by definition, two equivalent equations have the same graph.

• *Checking the equivalence*

Then ask students how they could use the idea that the equations should be equivalent to check their work. Help them see that if they take any solution to a "$y =$" equation and substitute it in the corresponding original equation, it should satisfy the original equation.

"Why is it relatively easy to get solutions to a 'y =' equation?"

Bring out that it's relatively easy to get solutions to a "$y =$" equation, because one can simply pick a value for x and substitute to get the corresponding value for y.

Have students try this for a couple of examples. For instance, they may have gotten the equation $y = \frac{17 - 5x}{3}$ as the "$y =$" form for the equation $5x + 3y = 17$ in Question 3. You might have different groups choose different values for x, substitute the values in the expression $\frac{17 - 5x}{3}$ to get y, and then check that their pair of numbers fits the original equation.

• *The issue of decimal approximation*

In solving an equation for y in terms of x, some students may use decimal approximations rather than fractions. For example, in Question 3, they may solve $5x + 3y = 17$ and get $y = 5.67 - 1.67x$ instead of $y = \frac{17 - 5x}{3}$.

If this comes up, you should raise the point that this may lead to slightly inaccurate graphs or approximate solutions.

For instance, if students substitute 2 for x in this decimal approximation, they get $y = 2.33$, and when they substitute into the original equation, $5x + 3y = 17$, they will get only an approximate solution; that is, $5 \cdot 2 + 3 \cdot 2.33$ comes out to 16.99 instead of 17. Thus, the equation $y = 5.67 - 1.67x$ is only *approximately equivalent* to $5x + 3y = 17$. Bring out that this inaccuracy may be unimportant in many contexts but important in others. The intent at this point is simply to alert students to the issue.

> *Note:* Some substitutions for x in the approximate equation do lead to exact solutions to the original equation. For instance, substituting 1 for x into $y = 5.67 - 1.67x$ gives $y = 4$, and the pair of numbers $x = 1$, $y = 4$ does solve $5x + 3y = 17$ exactly.

2. Solving *Homework 12: Rock 'n' Rap* on the Graphing Calculator

Yesterday, students saw that the solution to *Homework 12: Rock 'n' Rap* was the point where the lines $15{,}000x + 12{,}000y = 150{,}000$ and $y = x$ intersected. They may have found the coordinates either by plotting pencil-and-paper graphs or by guess-and-check. Today, students will see how to use graphing calculators, not only to find these coordinates but also to get graphical confirmation of how the "family of parallel lines" reasoning works.

"How can you make a calculator graph of the feasible region?"

Have students begin by graphing the lines that make up the boundaries of the feasible region. They will first have to rewrite the equations $15{,}000x + 12{,}000y = 150{,}000$ and $18x + 25y = 175$ in "$y =$" form in order to enter them on the graphing calculator.

They should then graph these rewritten equations and the equation $y = x$, in order to see the feasible region. Adjustments in the viewing rectangle may be needed.

"What do the profit lines look like?"

Next, have students look at the parallel lines they get for different profits. Again, they will have to rewrite the equation in "$y =$" form. For example, the profit equation $20{,}000x + 30{,}000y = 120{,}000$ becomes $y = (120{,}000 - 20{,}000x)/30{,}000$ (or something equivalent). But they now have the opportunity to check out various profit lines easily by changing the number 120,000 in the profit line equation and seeing what the new line looks like.

Let students play with this for a while, varying the profit and seeing how the profit line moves. This should reinforce the idea that the point where the profit is greatest is at the intersection of the two lines $y = x$ and $15{,}000x + 12{,}000y = 150{,}000$. Then they can use the trace feature (perhaps combined with adjustments in the viewing rectangle) to get the coordinates of the desired point. (We recommend that students use the trace feature, rather than an "intersect" or "solve" feature, because using the trace feature gives a more visual sense of what the coordinates mean. Once students understand this clearly, they will likely find shortcuts on their own.)

When students are done with the *Homework 12: Rock 'n' Rap* problem, you can ask them to work on *A Rock 'n' Rap Variation,* in which the profits are reversed.

You may want to have your more ambitious and independent students see if it is possible to *shade* the feasible region on the graphing calculator. They can consult the graphing calculator manual for details.

Homework 14: Going Out for Lunch

This homework involves a standard "two equations in two unknowns" problem. Tomorrow, students will be learning to solve such problems algebraically, but for now, they probably can solve this problem simply by guessing.

We don't want students to lose the skill of solving simultaneous equations more intuitively. Often, when students learn a mechanical way to solve problems, they preempt their intuition and end up not being able to solve the problem at all. Therefore, we would prefer at this time that they *not* focus on solving this problem with equations. (If some of them do it that way, it's okay, but don't include that approach in the homework discussion.)

Student Lidia Murillo can't decide which cookie to have for lunch.

Homework 14 Going Out for Lunch

Imagine! You have just started a full-time summer job in an office. It's your first day on the job, and the boss has sent you out to buy lunch for the 23 people who work in the office.

It turns out that everyone wants either one hot dog or one hamburger for lunch. You get to Enrico's Express, and Enrico asks you how many hot dogs and how many hamburgers you want. You realize you were so excited that you forgot to write down how many of each you were supposed to get. But you see that hamburgers cost $1.50 each and hot dogs cost $1.10 each (tax included), and that you were given $32.10.

Your Task

Assume that the $32.10 is the exact amount needed for your purchase.

1. Figure out, *in any way you can,* how many hot dogs and how many hamburgers were ordered.
2. Do you think the answer you found is the only one possible? Explain why or why not.

Adapted from *Algebra I,* by Paul A. Foerster, Addison-Wesley Publishing Co., 1990, p. 333.

Points of Intersection

This page in the student book introduces Days 15 through 18.

The feasible region for a system of inequalities gives you a picture of the possible options, and the family of parallel lines helps you see geometrically where to maximize or minimize a linear expression.

The next step is finding the exact coordinates of that maximum or minimum point. Often, that means finding a common solution for a pair of linear equations. In the main activity for the next few days, *Get the Point,* you will examine pairs of linear equations and develop one or more methods for finding their common solutions.

Erik Braswell and Casey Kelley will be able to explain several methods for solving systems of equations after meeting the challenge of "Get the Point."

Students begin developing an algebraic method for solving systems of equations.

Mathematical Topics

- Expressing a word problem as a system of two equations in two unknowns
- Beginning to develop an algebraic method for solving systems of linear equations

Outline of the Day

In Class

1. Form new groups
2. Discuss *Homework 14: Going Out for Lunch*
   - Set up the problem as a system of linear equations with carefully defined variables
3. Systems of linear equations
   - Introduce the term **system of equations**
   - Introduce the synonyms **solution to the system** and **common solution**
   - Relate linear systems to the unit problem
   - Solve *Homework 14: Going Out for Lunch* graphically
4. *Get the Point*
   - Students begin to develop an algorithm for solving simultaneous systems of equations algebraically
   - (Optional) Bring the class together to discuss Question 1a
   - The activity will be continued on Day 16, with presentations on Day 17

At Home

Homework 15: Only One Variable

Discuss With Your Colleagues

Why Make Up a Method?

In the activity *Get the Point,* students spend considerable time making up their own algorithms to solve systems of equations. Wouldn't it be faster to just tell them how to solve systems? Why might it be worthwhile to have students develop their own algorithm?

1. Forming New Groups

This is an excellent time to place the students in new random groups. Follow the procedure described in the IMP *Teaching Handbook* and record the groups and the suit for each student.

2. Discussion of *Homework 14: Going Out for Lunch*

"How would you solve this problem* without *equations?"

Have students share as many ways as possible of solving the problem without equations. If students begin to suggest a solution using equations, ask them to hold off.

"Is this the only solution? Why do you think so?"

Ask students whether they think there is more than one answer, and if so, what makes them think so. One possible intuitive explanation is that the more hot dogs there are, the less it costs for 23 people, and the more hamburgers there are, the more it costs. This means there cannot be more than one solution.

Whether students use this particular argument or another, try to get them to express their explanation as clearly as possible, and bring out that such an argument is, in fact, a proof of uniqueness.

• *Using equations*

"Now, how would you express this problem algebraically?"

After this intuitive discussion, ask students how they would express the problem algebraically. If a hint is needed, suggest that they define two variables and write equations expressing the information in the problem.

Insist that students define their variables clearly. For instance, do not accept statements like "x = hot dogs," because students may then confuse "*the number of* hot dogs" with "*the cost of* hot dogs." Point out that these are two separate numbers. Emphasize that when we use variables to represent unknown numbers, we have to state exactly what unknown numbers these variables stand for. Also, tell students that we often use the word *unknown* as a synonym in this context for a variable.

If students use x for the number of hot dogs and y for the number of hamburgers, then the equations might look like this:

$$x + y = 23$$

$$1.10x + 1.50y = 32.10$$

Once the equations have been presented, have students check that the solution arrived at informally satisfies them. Make sure they see that verification involves nothing more than substituting 6 for x and 17 for y into the equations and seeing that this gives true statements.

Do not take time discussing how to solve this pair of equations algebraically. Students will begin learning about that in today's activity, *Get the Point.*

3. Systems of Linear Equations

"What type of equations are these?"

Ask students what type of equations they used for the homework problem. If a hint is needed, you can ask how they would describe the expressions $x + y$ and $1.10x + 1.50y$ (or whatever they used), so that they identify the homework equations as *linear equations*.

Tell students that when they have a set of equations using the same variables and are looking for values for those variables that satisfy *all* the equations, we call this a **system of equations** (or system of *simultaneous* equations). A set of values that fits all the equations is called a **solution** (or *common solution*) **to the system.**

Tell them that in particular, the pair of equations they developed for *Homework 14: Going Out for Lunch,* is called a **system of linear equations.** (In the case of the homework problem, there are only two equations in the system, so we might call it a *pair* of linear equations.)

Tell students that in the activity they begin today, *Get the Point,* they will learn how to use algebraic methods to find the common solution to such a pair of linear equations.

• *Connection to the unit problem*

"How is the task of solving a system of linear equations related to the unit problem?"

Ask how the task of solving a system of linear equations is related to the unit problem. If a hint is needed, ask what type of equation defined the boundaries of the feasible region for that problem (and for others, such as *Homework 12: Rock 'n' Rap*).

You might remind students that in several problems they have seen in this unit, they needed to find the point where two linear graphs met. Bring out that finding such a point is the same as solving a system of linear equations.

• *Two types of problems*

Later in this unit, students will be developing their own "two-equation, two-unknown" problems (in *Homework 21: Inventing Problems*) and then in groups developing linear programming problems (in *Producing Programming Problems,* introduced on Day 23). You can start laying the groundwork now for the distinction between these two types of problems.

"What is the difference between 'Homework 14: Going Out for Lunch' and the unit problem?"

Ask about the difference between *Homework 14: Going Out for Lunch* and the unit problem. Without getting too formal, you can bring out that in *Homework 14: Going Out for Lunch,* the task is simply to solve a system of linear equations. By contrast, in the unit problem, the main task is to find a point where a certain linear expression (the *profit function*) is a maximum, and solving a system of linear equations is only a part of that task.

• *Solving with a calculator graph*

Point out that students used essentially a guess-and-check approach in solving the homework problem. Ask what other method they have seen for finding the solution to a system of linear equations. If necessary, remind students of their work yesterday with *Homework 12: Rock 'n' Rap* to bring out the idea of a graphical solution.

Ask students to graph the two equations from *Homework 14: Going Out for Lunch* on the graphing calculator to see if this also seems to give the same solution. They will need to solve the equations for y in terms of x (changing variables if they used something other than x and y). They will probably leave the second equation in a form like $y = (32.10 - 1.10x)/1.50$ rather than divide through. This should confirm their solution in yet another way.

4. *Get the Point*

The goal of the next activity, *Get the Point,* is for students to develop an algebraic method for solving pairs of linear equations. One approach is to solve both equations for one variable to get expressions in terms of the other variable, and then set those two expressions equal. Another approach is to solve one equation for one variable and then substitute the resulting expression for that variable in the other equation.

Students may even find that they like one method for some systems and another method for other systems. The goal is not for them to develop any particular method, but for them to find an algebraic procedure (or several such procedures) that works and makes sense to them.

Note: The introductory paragraphs of this activity continue to reflect the dual graph/equation approach that we are emphasizing.

Tell students that in the next activity, *Get the Point,* they will develop some systematic algebraic way (or ways) to find the exact solution to a system of two linear equations. You may want to emphasize that the algebraic approach will always result in the exact solution, which gives it an advantage over graphing.

Students will work on this activity the rest of today and all of tomorrow (except for discussion of tonight's homework). Presentations are scheduled for Day 17.

Emphasize that students will produce individual written reports on this activity and that groups will make oral presentations. The reports are scheduled to be completed tomorrow, with presentations to be done the following day. (You may need to modify this timetable as you observe your students' work. You should review the timetable with them so they know what to expect.)

Here are some hints that you might use either today or tomorrow as students work on this activity.

• *Hints on Questions 1a and 1b*

As students work on Question 1, you may need to give hints to groups to help them develop their method. But let groups wrestle with it for a while before you intervene.

Possible hints: "What is true of all points on the line $y = 3x$?" or "How would you get the y-coordinate from the x-coordinate for a point on this line?"

Questions 1a and 1b represent the simplest case, because they involve equations with y already explicitly stated in terms of x. If students need a hint for Question 1a, you can ask, "What is true of all points on the line $y = 3x$?" or "How would you get the y-coordinate from the x-coordinate for a point on this line?" Try to elicit an answer to the effect that the y-coordinate is three times the x-coordinate for all points on the line.

"Which of the two expressions, $3x$ or $2x + 5$, should you use to compute the y-coordinate at the point of intersection?"

You can then ask a similar sequence of questions for the line $y = 2x + 5$ and relate the two equations by asking what happens at the point of intersection. For instance, you might ask, "Which of the two expressions, $3x$ or $2x + 5$, should you use to compute the y-coordinate at the point of intersection?" Students should recognize that it doesn't matter. You can then bring out that students can find the x-coordinate by figuring out what value of x makes the two expressions equal, that is, by solving the equation $3x = 2x + 5$.

A similar approach will work for Question 1b.

• *Whole-class discussion of Questions 1a and 1b*

You can have one or two of the faster groups prepare presentations on Questions 1a and 1b. When most groups seem to have made some progress on the two questions, bring the class together for a preliminary discussion. It may be helpful to have presenters graph the equations on an overhead graphing calculator.

Once the discussion has gone as far as it can productively, have students continue work on the activity.

• *Hints on Questions 1c through 1e*

Some groups may be clear on Questions 1a and 1b but not know what to do with the more complex problems. You can give some hints as you circulate, but again, be sure to give groups plenty of opportunity to develop the ideas on their own. Keep in mind that they have both Day 15 and Day 16 to work on this activity, so they have some time to grapple with the problems before they achieve success.

As hints for Question 1c: "Why is this pair of equations harder than the pair in Question 1b? How could you change the problem so it would be similar?"

If you do need to give hints, you might ask, "Why is the pair of equations in Question 1c harder than the pair in Question 1b? How could you change the problem so it would be similar?" Focus students' attention on the idea of solving the first equation for y.

The terminology of "equivalent equations" might also help groups that are struggling.

• *Question 2*

The basic goal on Question 2 is for students to articulate a general outline of the method of solution.

For groups that finish earlier than others, you can suggest that they think about finding a pair of equations whose graphs intersect in exactly two points, exactly three points, and so on. Point out, if necessary, that this will involve nonlinear equations. Of course, this is an enormous question, and we don't expect students to get any systematic or comprehensive answers.

• *Two possible outlines*

One possible outline for Question 2 might look like this. You might refer to this as the "setting y's equal" method.

1. Use each of the equations to get an expression for y in terms of x.
2. Set these two expressions equal to each other.
3. Solve the new equation you just created, which has only one variable, x.
4. When you find x, substitute it in one of the original equations and solve for y.

Groups using this general approach will probably vary as to how much detail they give on step 1 or step 3. Do not expect a complete algorithmic procedure for step 3. In fact, even doing so for step 1 will be difficult.

Other groups may use quite different methods, and you should encourage variety. For instance, here is another possible outline. This method is usually called **substitution.**

1. Solve one of the equations to get an expression for y in terms of x.
2. Replace y in the other equation with the expression from step 1.
3. Solve the new equation you just created, which has only one variable, x.
4. When you find x, substitute it in one of the original equations and solve for y.

As noted earlier, some groups may choose to describe more than one method and may suggest using one method in some cases and a different method in others.

- ***For teachers: Why not the elimination method?***

For many of us, the preferred method of solving systems of equations in several unknowns is the "elimination" method, in which you multiply equations by constants and then add or subtract equations to eliminate one variable at a time. Although the elimination method is often faster than the other methods just described and might make it easier to avoid fractions, we are not recommending the elimination method now. Why not?

The basic reason is that the approach of solving both equations for y and setting the expressions equal to each other is easier to conceptualize than the elimination method. Students already see the idea of solving for y in terms of x as an important step in graphing on the calculator, and the approach we are suggesting here builds on that technique.

We want students to have at least *one* method that they can rely on fully, understand completely, and fall back on in case they need to. If students do individual problems by another method or discover elimination on their own, that's fine, but we do not want to present it now.

In the Year 3 unit *Meadows or Malls?* students will learn the elimination method and ways to work with more than two variables. They will also learn about matrices and discover how to use them to express systems of linear equations. They will then learn how to use graphing calculators to invert matrices and thus have the calculators do all the tedium for them.

Homework 15: Only One Variable

Tonight's homework is designed to help students review ideas from *Solve It!* about the solution of linear equations in one variable.

Get the Point

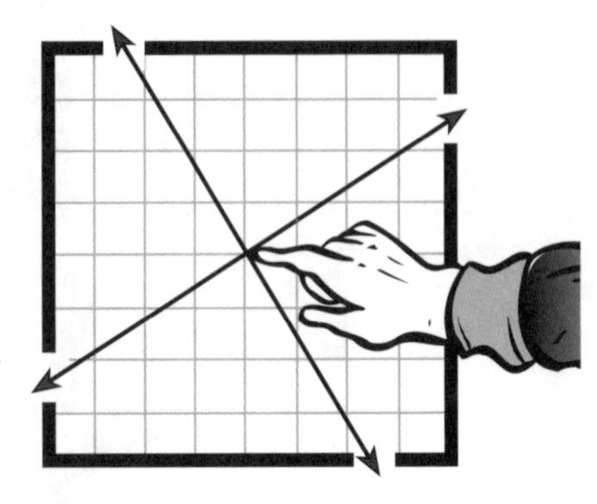

In solving problems like the cookie problem, it is helpful to know how to find the coordinates of the point where two lines intersect. As you have seen, this is equivalent to finding the common solution to a system of two linear equations with two variables.

You have probably done this already using either guess-and-check or graphing. Your goal in this activity is to develop an algebraic method, by working with the equations of the two straight lines.

Your written report on this activity should include two things.

- Solutions to Questions 1a through 1e
- The written directions your group develops for Question 2

1. For each of these pairs of equations, find the point of intersection of their graphs by a method other than graphing or guess-and-check. When you think you have each solution, check it by graphing or by substituting the values into the pair of equations.

   a. $y = 3x$ and $y = 2x + 5$

   b $y = 4x + 5$ and $y = 3x - 7$

   c. $2x + 3y = 13$ and $y = 4x + 1$

   d. $7x - 3y = 31$ and $y - 5 = 3x$

   e. $4x - 3y = -2$ and $2y + 3 = 3x$

2. As a group, develop and write down general directions for finding the coordinates of the point of intersection of two equations for straight lines using an algebraic method, without guessing or graphing. In developing these instructions, you may want to make up some more examples like those in Question 1, either to get ideas or to test whether your instructions work.

   Make your instructions easy to follow so someone else could use them to "get the point."

Homework 15 Only One Variable

In *Get the Point,* your goal is to develop a method for solving systems of two linear equations in two variables. In *Solve It!* you saw that you could solve linear equations in one variable, and you may find it helpful to review that process as you work on *Get the Point.*

1. Solve each of these linear equations.

   a. $5x + 7 = 24 - 6x$

   b. $6(x - 2) + 5x = 9x - 2(4 - 3x)$

   c. $\frac{x + 3}{2} = 29 - 2x$

   d. $\frac{3x + 1}{4} = \frac{5 - 2x}{6}$

2. Make up a real-world question that can be represented by a linear equation. Try to create an example in which the variable will appear on both sides of the equation.

DAY 16 *Getting the Point Some More*

Students continue to work on developing an algebraic method of solution.

Mathematical Topics

- Continuing to develop an algebraic method for solving a pair of linear equations
- Simplifying equations involving fractional expressions

Outline of the Day

In Class

1. Discuss *Homework 15: Only One Variable*
   - Review techniques and concepts for manipulating expressions and equations
2. Continue *Get the Point!* (from Day 15)
   - Students continue developing an algorithm
   - Students should prepare their reports for tomorrow's presentations

At Home

Homework 16: Set It Up

1. Discussion of *Homework 15: Only One Variable*

You might let students discuss the homework problems briefly in their groups and then have club card students present their work. For each problem, you can follow up by asking for alternative approaches.

Question 1a should be straightforward. You can use the presentation to review the basic principles for getting equivalent equations, perhaps referring to the mystery bags game of *Solve It!*

If needed, you can use Question 1b as a lead-in for a review of the distributive property and methods for removing parentheses. Having students check the solution (by substitution) will also give you an opportunity to review the arithmetic of integers.

- *Questions 1c and 1d*

Questions 1c and 1d are designed to anticipate problems students may have in their continued work on *Get the Point* later today. Specifically, you can take this opportunity to discuss how to work with equations involving fractions.

On Question 1c, some students might rewrite the equation as $0.5x + 1.5 = 29 - 2x$, which allows them to avoid the issue of fractions. Others might multiply both sides by 2 and get $x + 3 = 2(29 - 2x)$. Still others might get the same result by viewing the fraction as division and using the idea that if $a \div b = c$, then $a = bc$. However the presenter solves this equation, try to bring out other approaches.

Question 1d is more complicated than Question 1c because fractions (with different denominators) appear on both sides of the equation. One standard approach is to eliminate the fractions by multiplying both sides first by 4 and then by 6. (Students may also see that they can eliminate fractions by multiplying directly by 24 or by 12, but you need not concern them with multiplying by the least common multiple.)

If students are comfortable with the idea of "cross multiplying" and can explain why it works, then that is also a good approach.

- *Question 2*

You can have two or three volunteers share the problems they created for Question 2. Then let the class try to write the equations that go with these problems.

2. Continuation of *Get the Point*

Have students continue their work on *Get the Point.* It may be appropriate to go over some key points as a class, depending on how much success groups had yesterday. Within their groups, students should be developing their general directions (Question 2). Before the day is finished, they should prepare their final reports.

One purpose of Question 2 is to sharpen students' sense of precision and to get them to articulate their ideas clearly. You might want to give groups some constructive criticism of their directions for Question 2 before they write them on transparencies. The outlines for Question 2, given in the Day 15 introduction to *Get the Point,* provide an idea of the overall expectations for this activity.

Homework 16: Set It Up

Question 1 of tonight's homework combines several components of students' work so far in this unit: setting up equations, solving equations using a graph, and solving equations algebraically.

Question 2 represents a twist on solving equations, in that students start with the solution and have to make up the pair of equations.

Before preparing his presentation to the class, John Idsten explains to his group how he set up his equations.

Homework 16 Set It Up

1. You now have more understanding of how to set up and solve pairs of equations in two variables. This problem gives you a chance to apply your knowledge.

   Marvelous Marilyn scored 273 points last season for her high school basketball team. Her points resulted from a combination of two-point shots and three-point shots. She made a total of 119 shots. How many of each type of shot did Marilyn make?

   In writing up this problem, follow these steps.

   - Choose variables and state what they represent.
   - Write a pair of equations using your variables that represent the problem.
   - Solve the pair of equations graphically.
   - Solve the pair of equations algebraically.
   - Answer the question in the problem.

2. Make up a pair of linear equations whose common solution is $x = 3$ and $y = 5$.

DAY 17

Presenting the Points

Students present reports on "Get the Point."

Mathematical Topics

- Articulating an algebraic method for solving a system of two linear equations in two variables

Outline of the Day

In Class

1. Select presenters for tomorrow's discussion of *POW 12: Kick It!*
2. Discuss *Homework 16: Set It Up*
   - Have students define their variables carefully
   - Discuss the relationship between algebraic and graphical methods of solution
   - Bring out that approximating fractions by decimals may lead to inexact answers
3. Presentations of *Get the Point*
   - Ask for different methods
   - Acknowledge that different examples may call for different approaches

At Home

Homework 17: A Reflection on Money

Discuss With Your Colleagues

Aren't Fractions More Precise?

In solving for one variable in terms of another, students often use decimals rather than fractions. If they do this, they might not get exact answers. Why not just use a graphing calculator if you want an approximate answer? Should you insist that students use fractions?

1. POW Presentation Preparation

Presentations of *POW 12: Kick It!* are scheduled for tomorrow. Choose three students to make POW presentations, and give them overhead transparencies and pens to take home to use in their preparations.

2. Discussion of *Homework 16: Set It Up*

You can give transparencies to different groups to prepare presentations on different parts of the assignment. Have spade card students do the presentations. You might have one group both define variables and give equations, another solve the equations algebraically, and a third solve the equations graphically. You can leave answering the question to the whole class.

You may want to have two or three presenters for Question 2, because there are many different linear systems that have the given solution.

Note: If you have identified groups that need a few more minutes to finish preparing their presentations on *Get the Point,* you can let them do this while other groups prepare the homework presentations.

- *Question 1*

The first presenter on Question 1 should have a pair of equations something like this:

$$x + y = 119$$

$$2x + 3y = 273$$

where x represents the number of two-point shots and y represents the number of three-point shots. Focus on careful definition of variables—for example, not "x = two-point shots," but "x = the number of two-point shots."

In the graphical approach, be sure to have the presenter explain how he or she went about sketching the graphs. This can be an opportunity to review ideas about graphing linear equations that may have been discussed in *Solve It!* such as use of intercepts as easy points to plot.

Similarly, with the presentation of an algebraic method, have the presenter give details on what he or she did. This presentation may give you some idea of what to expect later today in the presentations on *Get the Point.*

- ### *The decimal approximation issue again*

In the Day 14 discussion of *Homework 13: Getting on Good Terms,* we raised the issue of student use of decimal approximations (see the subsection "The issue of decimal approximation"). That issue may arise again here, because an approximate expression for y in terms of x may lead to only an approximate solution to the system of equations.

For instance, suppose students rewrite the equation $x + y = 119$ as $y = 119 - x$ and rewrite $2x + 3y = 273$ first as $3y = 273 - 2x$ and then as $y = 91 - 0.67x$.

They may then set the two expressions for y equal to each other, which gives

$$119 - x = 91 - 0.67x$$

If they combine terms to get $28 = 0.33x$ and then divide both sides by 0.33, they will get $x \approx 84.85$. Substituting this value for x in the equation $x + y = 119$ gives $y \approx 34.15$. This is only an approximation to the exact answer, which is $x = 84, y = 35$.

If this issue arises, caution students that they should test their answer in both original equations. For example, if they substitute 84.85 for x and 34.15 for y into the expression $2x + 3y$, they would get 272.15, and not 273 as required.

This situation should be a warning to them about losing exactness when they round off in decimals.

- ### *Back to the problem*

After presentation of both graphical and algebraic methods for solution of the system, ask for a volunteer to answer the original question in the problem. You can use this as an occasion to emphasize that the algebra and the graph are simply tools for solving the problem; it's the problem itself that is the final arbiter of the correctness of the process.

- ### *Question 2*

"How did you find the system?"

Have students who prepared presentations for Question 2 show their work. Ask them to explain how they found the system.

If no one describes the method of simply starting with a linear expression in x and y and substituting to get the constant term, you might ask something like, "Is there an equation you could use that has $7x + 9y$ on the left side?"

"How can you explain in terms of graphs why there are different systems with this solution?"

Presumably, different students will have come up with different systems with the given solution. Raise the question of why there is more than one possible answer, specifically asking for a graphical explanation.

Students should see that they can use any pair of equations whose graphs are distinct lines through the point with coordinates $x = 3$, $y = 5$. Also, they should bring out that there are infinitely many lines through this point, any pair of which will work for Question 2.

3. Presentations of *Get the Point*

Have several heart card students each read the procedure that their group developed for finding the coordinates of the point of intersection.

Throughout the presentations and discussion, the emphasis should be on using a method that works in the given situation and that makes sense to the user. Try to avoid giving the impression that there is a "best" method or that students need to learn a specific method and stick with it all the time.

"Is there anything about this procedure that isn't clear? How could it be clarified?"

As each procedure is being read, have other students make note of any aspects of it that are not clear. Then spend time going over each set of directions, with students suggesting ways of clarifying the statement. Have students test these procedures by working on specific problems.

Different groups will probably have developed sets of directions that vary at least in the details, if not in the broad outline.

- *Decimals and fractions*

If students have been converting fractions to decimal approximations, this is a good time to have them at least see how to do such problems in terms of fractions.

For instance, with Question 1c, they should be able to solve $2x + 3y = 13$ for y using fractions, writing $y = \frac{13 - 2x}{3}$ and setting $\frac{13 - 2x}{3}$ equal to the expression $4x + 1$ (from the second equation, $y = 4x + 1$). If students aren't sure how to solve the equation $\frac{13 - 2x}{3} = 4x + 1$ without using decimals, you may want to have them work on this in groups. As a hint, you can ask them to summarize the basic methods for getting equivalent equations. Mentioning the idea of multiplying both sides by the same factor should be a sufficient hint for some groups. With this hint, they should be able to rewrite $\frac{13 - 2x}{3} = 4x + 1$ as $13 - 2x = 3(4x + 1)$ and work from that equation to get an exact solution.

If students approximate $\frac{13 - 2x}{3}$ as $4.33 - 0.67x$ and then solve the equation $4.33 - 0.67x = 4x + 1$, they will probably get $x = \frac{3.33}{4.67}$, which is approximately 0.713. Bring out that the exact solution for x, which is $\frac{10}{14}$ (or $\frac{5}{7}$), comes out to approximately 0.714. Although the difference is small, you should emphasize that in some situations, it may be important to obtain the exact value. (You can refer students to *Homework 19: Falling Bridges* in *Do Bees Build It Best?* as an illustration of this issue.)

- *Variations from problem to problem*

As noted before, students may find that different approaches work better on different problems. For instance, on Question 1c, the "setting y's equal" method essentially involves first dividing by 3 (to solve the first equation for y) and then multiplying by 3 (after setting the expressions equal). But the substitution method eliminates these two steps by simply replacing y in the first equation with the expression $4x + 1$ from the second equation. This approach gives $2x + 3(4x + 1) = 13$, thus avoiding fractions completely.

On the other hand, in Question 1e, there is no way to get one variable in terms of the other without introducing fractions or decimals, and substitution is probably not any easier.

If you use the "setting y's equal" method in Question 1e, you get

$$\frac{4x + 2}{3} = \frac{3x - 3}{2}$$

If you use the substitution approach, solving the equation $2y + 3 = 3x$ for y and substituting in the other equation, you get

$$4x - 3\left(\frac{3x - 3}{2}\right) = -2$$

Probably, most students would find the first of these two equations easier to work with.

Students may come up with shortcuts on specific problems they create. For example, in the pair of equations $6x - 7y = 47$ and $2x + 5y = -21$, it's convenient to solve the second equation for $2x$ and use the fact that $6x = 3(2x)$ to substitute in the first equation.

Homework 17: A Reflection on Money

Tonight's assignment continues the work with word problems, systems of linear equations, and method of solution. It also gives students an opportunity to reflect on the graphical and algebraic approaches.

Homework 17 A Reflection on Money

1. Read the problem below about Uncle Ralph. You are asked to solve it in two different ways and then reflect on your method of solution.

   Uncle Ralph says that if you can tell him the number of each type of coin in his pocket, then you can have the money. He gives you this information.

   - He has only dimes and quarters.
   - He has 17 coins in his pocket.
   - The coins are worth \$3.35.

   Can you get Uncle Ralph's money?

   a. Solve the problem graphically.

   b. Solve the problem algebraically.

   c. Describe the advantages and disadvantages of the graphical and algebraic methods.

2. Here are three systems of linear equations. Solve each system algebraically, using either the method you developed in *Get the Point* or any other methods you have learned. Explain your work clearly.

   a. $y = 2x - 3$ and $3x - 4y = 7$

   b. $c + 2f = -6$ and $3c + f = 2$

   c. $2r - k = -1$ and $6r = 5k - 11$

DAY 18

POW 12 Presentations

Presenters give their conclusions on POW 12 and students share POW write-ups.

Mathematical Topics

- Comparing the graphical and algebraic methods of solving equations
- Developing relationships between linear combinations of integers and their common divisors

Outline of the Day

In Class

1. Discuss *Homework 17: A Reflection on Money*
   - Discuss the advantages and disadvantages of the graphical and algebraic methods
2. Presentations of *POW 12: Kick It!*
   - Have presentations, but do not ask for ideas about the POW from other students
   - List conclusions proposed by the presenters, and have the class label them as either "proven conclusion" or "conjecture"
   - Students read and comment on one another's POWs, and can improve their write-ups tonight

At Home

Homework 18: More Linear Systems

POW 13: Shuttling Around (due Day 24)

Discuss With Your Colleagues

Why the Emphasis on Graphs?

Students now have an algebraic method for finding the coordinates of the point of intersection of two lines. Nevertheless, they were asked in *Homework 17: A Reflection on Money* to solve a system graphically. Why is it important to continue using graphs?

1. Discussion of *Homework 17: A Reflection on Money*

Have students discuss the homework in groups. Solving the problem in Question 1 and the systems in Question 2 may be fairly routine by now, although the systems in Question 2 do have negative and fractional answers. Use your judgment about whether you need to have presentations on either of these questions.

Note: Because the examples in *Get the Point* all used x and y, some students may have been confused by the presence of different variables in Questions 2b and 2c. You might use the presence of other variables to bring out that for the purpose of getting an algebraic solution, one can solve for either variable in terms of the other, even if the variables are x and y. (If students have an equation using x and y and want to graph it on the calculator, they do need to solve for y in terms of x.)

If there do not seem to be difficulties, you can focus the discussion on Question 1c, perhaps asking several diamond card students to report on their ideas about the advantages and disadvantages of the graphical and algebraic approaches.

2. Presentations of *POW 12: Kick It!*

Ask the three students selected yesterday to make their presentations. Presumably, on Question 1, they will have found that the highest impossible score is 7. The focus here should be on their reasoning as to why every higher score is possible, and the quality of their explanations may vary.

If the presenters are having difficulty developing a convincing explanation, you can suggest that they get a sequence of consecutive scores that are possible and then see how every larger score can be obtained from a score in that sequence.

For example, students might show explicitly that all scores from 10 through 19 are possible and then observe that they can obtain any larger score by adding 10 points (2 field goals) as often as necessary to one of these scores. For instance, they can get a score of 47 by taking the scoring combination that gives 17 and adding 6 additional field goals.

Note: More "efficient" variations of this proof are possible, but using multiples of 10 makes this one comparatively easy for students to see.

- ***Questions 2 and 3***

As presenters discuss Questions 2 and 3, they will be offering conclusions either in connection with specific other scoring systems or as general principles. Have the class make a list of these conclusions as they are presented, assigning them to one of two sheets of chart paper, which you might label "Proven Conclusions" and "Conjectures." For example, if the presenter gives a proof of a statement—that is, an explanation showing that the conclusion is true—list it on the sheet labeled "Proven Conclusions." If the presenter offers a conclusion without satisfactory proof, list it on the sheet labeled "Conjectures."

You can let the listening students take the lead in deciding which statement goes under which label. However, you may need to intervene if students think that an incorrect statement has been proved or if you believe the explanation is not sufficient. When in doubt, it's probably better to err on the side of caution, labeling the statement "Conjecture."

Save these two lists so students can work with them tomorrow.

In tomorrow's continuation of the discussion of this POW, you may end up moving some of the conjectures either to the "Proven Conclusions" list or to a new list, labeled "Disproved Conjectures." Tell the class that they should consider statements on the "Conjectures" list undecided for now, and suggest that they think further about whether these are true. Inform them that they will continue work with these lists tomorrow.

- ***Students read each other's papers***

Note: Usually, after POW presenters are done, you ask other students to add further ideas. In this case, we are suggesting a different approach in which students read one another's papers and continue the discussion tomorrow.

When the three presenters are done, have groups exchange their papers with other groups, read these other papers, and note places where the write-ups are unclear or incomplete. (If you have students who have not done a POW write-up, you can have them work on the POW at this time.) Then have them return the papers.

Tell students they will rewrite their POW tonight, based on the written feedback and on what they learned from the presentations. Tomorrow, they will hand in their first version, with the comments other students made, and their new version.

Note: There is no written assignment in the student materials regarding this POW revision. We recommend that this revision be done in addition to *Homework 18: More Linear Systems,* which is fairly routine.

- *Sample conclusions and proofs*

Here are some examples of conclusions and proofs that might arise. We urge you not to read these until you have worked on the problem yourself. We also caution that these ideas are intended as background information for you, and not as a set of expectations of what students should do.

Example 1: The first example concerns a case in which there is no highest impossible score.

Conclusion: If the number of points for a field goal is 2 and the number of points for a touchdown is 8, then there is no highest impossible score.

Proof: You can get only even scores in this game, so every odd number is impossible. Because there is no highest odd number, there is no highest impossible score.

Note: You may get several conclusions of this type. The familiar terminology of *even* and *odd* makes this example easier to state than a case in which the scores are, for example, 5 and 15.

If you get several cases like this, you may want to try to move students toward this more general statement of the conclusion:

Conclusion: If the number of points for a field goal is more than 1 and divides the number of points for a touchdown, then there is no highest impossible score.

Proof: All the scores made only with field goals are multiples of the points for one field goal. Because the points for each touchdown are also multiples of that number, then all possible combinations of the two types of score are multiples of that number. Therefore, every number that is not a multiple is impossible. Because we can always find a number larger than any given one that isn't a multiple, there is no highest impossible score.

Note: This conclusion is stated in terms of the score for a field goal dividing the score for a touchdown, but it would work just as well the other way around. See "For teachers: Further background information" for an even better generalization.

Example 2: The next example concerns a family of cases in which there *is* a highest impossible score. The reasoning is similar to that for the specific case mentioned in the POW.

Conclusion: If the number of points for a field goal is 2 and the number of points for a touchdown is odd, then the highest impossible score is 2 less than the score for a touchdown.

Proof: You can get any even score from field goals and any odd score at or above the value of a touchdown by using one touchdown and the right number of field goals. But you can't get any odd numbers less than the value of a touchdown.

- ***For teachers: Further background information***

The following observations also apply to this POW. This material is presented as further background information, not as material that should be presented to the class. A POW investigation is intended to open doors for students, but not as a topic that students need to master.

You may want to use some of the ideas below as topics for students to pursue for extra credit or simply for the fun of further exploration.

a. Whenever the two numbers have a common factor bigger than 1, all scores will be multiples of that common factor, and so there will be arbitrarily high impossible scores. This is an even better generalization of Example 1 than the one given earlier. (For example, it includes a case like 6 points for a field goal and 8 points for a touchdown.) Explaining this should be within students' grasp.

b. Conversely, if the two numbers are *relatively prime* (that is, have no common factor except 1), then there will be a highest impossible score. However, proving this in general is difficult, and probably beyond the experience of students.

c. A general rule for finding the highest impossible score is

$$ab - (a + b)$$

where a and b are the two values and a and b are relatively prime. Some students may find this formula through examples, but the proof is difficult. Students also may find formulas for special "families." For example, as indicated in Example 2, if a field goal is worth 2 points and a touchdown is worth an odd number of points, then the highest impossible score is 2 less than the value of a touchdown.

d. As an extension, students might consider the situation of more than two kinds of scores—for example, field goals are worth 6 points, touchdowns 10 points, and safeties 15 points. (The numbers in this particular example have the peculiar property that any two of them have a common factor greater than 1, but no integer greater than 1 divides all three.)

POW 13: Shuttling Around

You should allow about 5 minutes to introduce the new POW. Perhaps the best way to begin the introduction is to let students work out the simplest case of the puzzle, which has one of each type of marker, as shown below.

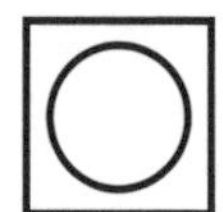  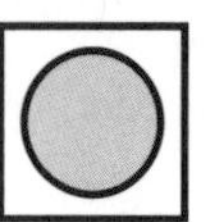

They should see quickly that the puzzle can be solved in three moves.

Then let them work in pairs for a few minutes on another example. Emphasize that a marker can jump over only one other marker.

This POW is scheduled for discussion on Day 24.

Homework 18: More Linear Systems

Tonight's assignment continues students' work with linear equations and includes both a one-variable equation with no solution and an inconsistent system in two variables. Let students discover the complications of these examples on their own and struggle at home with how to handle them. (They saw one-variable equations with no solution in ***Solve It!***) Tomorrow's discussion will deal with both inconsistent and dependent systems of linear equations.

Vanessa Ayala and Monica Sanchez appear pleased with the "family of parallel lines" displayed on their calculator.

POW 13 *Shuttling Around*

This POW is about solving a puzzle or, rather, about solving a whole set of puzzles. Each of these puzzles requires two sets of markers, such as coins of two different types. We will use plain and shaded circles to represent the markers.

An Example

One of these puzzles uses three markers of each kind. At the beginning of the puzzle, the markers are arranged as shown below, with each marker in a square. The plain markers are at the left, the shaded markers are at the right, and there is one empty square in the middle.

     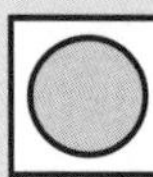 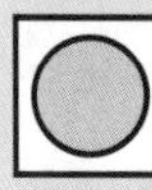

The task in the puzzle is to move the markers so that the shaded markers end up at the left and the plain markers end up at the right. Of course, there are some rules.

- The plain markers move only to the right and the shaded markers move only to the left.
- A marker can move to an adjacent open square.
- A marker can jump over *one* marker of the other type into an open square.
- No other types of moves are permitted.

Your Task

The reason that this is a *set* of puzzles is that you can vary the number of markers. Your POW is to investigate this set of puzzles. Begin with the example above, and answer these questions.

1. Can the puzzle be solved? If so, can you find more than one solution?
2. If the puzzle can be solved, how many moves are required? Is there a minimum? Can you prove your answer?

Continued on next page

Once you have answered these questions for the case in which there are three markers of each type, look at other examples. Consider only cases in which the numbers of each type are equal and there is exactly one empty square in the middle. (The supplemental problem *Shuttling Variations* examines other cases.)

Here are some things you can do.

- Find out if all such puzzles have solutions, and if so, how many moves are required.
- Look for a rule that describes the minimum number of moves in terms of the number of markers.
- Prove your results.

Write-up

1. *Problem Statement*
2. *Process:* Describe how you investigated this set of puzzles. Which cases did you examine? Did you try to develop any general conclusions?
3. *Conclusions and Conjectures:* State the questions you investigated and the conclusions you reached. Include questions that you did not have time to investigate. Some of your conclusions may be about specific cases while others may be general.

   If you can prove any of your conclusions, include the proofs. If any of your conclusions are still tentative, label them as conjectures.
4. *Evaluation*
5. *Self-assessment*

Homework 18 More Linear Systems

This assignment continues the work with linear equations. Keep alert for new shortcuts and new insights into how to solve either one-variable or two-variable equations.

1. Find the solution to each of these linear equations.

   a. $3(c + 4) - 2c = 16 - 4(c + 5)$

   b. $t + 2(t - 4) = 5(1 - 2t)$

   c. $\frac{r + 5}{2} = 12 - 3r$

   d. $7w + 2(3 - 2w) = 4(w + 2) - (w - 6)$

2. Find the value of both variables in each of these linear systems.

   a. $4a - 5b = -4$ and $3a + 6b = 10$

   b. $u - v = 3$ and $2u + 2v = 10$

   c. $2x + 3y = 1$ and $6y = 7 - 4x$

Cookies and the University

You're now ready to solve the unit problem. Use the feasible region, a family of parallel profit lines, and a pair of linear equations to find the cookie combination that the Woos are looking for.

When you're done with that, you'll apply the ideas to a completely new problem about college admissions.

This page in the student book introduces Days 19 through 21.

Vanessa Morales, Mandy Mazik, Anandika Muni, and Moses Lazo work together to solve the unit problem.

DAY 19 Back to Cookies

Students learn that not all systems of equations have exactly one solution and begin the final steps of solving the unit problem.

Mathematical Topics

- Solving systems of linear equations algebraically
- Seeing that a system of two linear equations in two unknowns may not have exactly one solution
- Introducing the concepts of inconsistent equations and dependent equations

Outline of the Day

In Class

1. Finish discussion of *POW 12: Kick It!*
   - Add to the lists begun yesterday for conclusions and conjectures
2. Discuss *Homework 18: More Linear Systems*
   - Bring out that a pair of linear equations may have no common solution
   - Relate the algebra of linear systems to the graphs of the equations
   - Introduce the terms **inconsistent equations**, **dependent equations**, and **independent equations**
3. *"How Many of Each Kind?" Revisited*
   - Students apply the concepts from the unit to the unit problem
   - The activity will be continued on Day 20, with presentations on Day 21

At Home

Homework 19: A Charity Rock

1. Conclusion to Discussion of *POW 12: Kick It!*

"Do you have any new conjectures or any comments on the old ones?"

Ask for volunteers to add to or comment on the two lists begun yesterday: "Proven Conclusions" and "Conjectures." There are several different ways in which students can contribute at this point.

- A student can offer a proof of a statement on the "Conjectures" list. If the class considers the proof convincing and you agree, move the statement to the "Proven Conclusions" list.
- A student can offer a counterexample to a statement on the "Conjectures" list. In this case, move the statement to a new list called "Disproved Conjectures."
- A student can offer a new statement, either with or without a proof. If the student presents a convincing proof (or if someone else can offer one), add it to the "Proven Conclusions" list. If someone has a counterexample to the proposed statement, put it on the new "Disproved Conjectures" list. If there is neither a proof nor a counterexample, add the statement to the "Conjectures" list.

After the discussion is completed, collect students' material on the POW, which will include their original write-ups, the comments they received from other students, and their revised write-ups.

2. Discussion of *Homework 18: More Linear Systems*

Let students compare answers from the homework. Then begin the discussion, perhaps by asking if students had any difficulties with or have any issues to raise about Question 1.

If no one points to Question 1d, bring it up yourself, and ask what students got for that problem. Because simplifying each side leads to the equation $3w + 6 = 3w + 14$, the equation has no solution, and students should recognize that fact. You might remind them that they saw a situation like this in *Solve It!* with the mystery bags game.

- *Question 2*

You can take a similar approach with Question 2, discussing Questions 2a and 2b only if students raise questions about them.

Ask for a volunteer to discuss Question 2c. (If no one feels confident about it, you can ask for someone to explain what was more difficult about it.) Probably the student will at least say that his or her usual method led to "something weird." You can probably connect that to Question 1d, which had no solution.

"What is happening in terms of the graphs of these equations? What usually happens with a pair of linear equations?"

If students are unable to explain what's happening in Question 2c, ask what is going on in terms of graphs. If needed, ask what finding the common solution to a pair of linear equations *usually* means in terms of graphs.

The response should bring out that we are usually looking for the point where two lines intersect, and this may give students some insight into the situation. If necessary, have them graph the two equations from Question 2c (which requires expressing y in terms of x in each case).

Bring out that the two lines are parallel and so have no points in common. Therefore the equations have no solution in common. Tell students that a pair of linear equations that have no common solution is called an **inconsistent system.**

• *Dependent systems*

"What else can happen with two lines (besides having a unique intersection or being parallel)?"

Point out that two lines "usually" will intersect in a unique point and that students have just seen a case of linear equations in which the two lines were parallel.

Note: Because there are infinitely many pairs of lines, it's difficult to define the probability that two lines will intersect in a unique point. The word "usually" is used here in a colloquial or intuitive way, and not in the more precise mathematical sense of "with probability greater than $\frac{1}{2}$."

If a hint is needed: "What about the pair of equations $2x + 3y = 1$ and $6y = 2 - 4x$?"

Ask if there is anything else that can happen with two lines besides having a unique intersection or being parallel. If a hint is needed, ask students about the pair of equations $2x + 3y = 1$ and $6y = 2 - 4x$. (These are the same as those in Question 2c except that the second constant term has been changed.)

Students should see (by solving for y and graphing, if needed) that these two equations have the same graph; that is, they are equivalent equations. You can point out that this might not have been obvious at a glance, because the equations have a somewhat different form. Bring out that the first equation is equivalent to $4x + 6y = 2$ (by multiplying it by 2) and that this is essentially the same as the second equation.

"Are there any number pairs that fit both of these equations?"

Ask if there are any number pairs that fit both of these equations. Students should see that any number pair that fits one will fit the other (because they are equivalent), so there are infinitely many solutions to the system. You might want to have students list two or three such common solutions.

Tell students that a pair of linear equations that have the same graph is called a **dependent system,** and that when two linear equations give distinct but intersecting lines, the equations are called **independent.**

- ***Three cases: Unique solution, inconsistent system, and dependent system***

It will probably be helpful to summarize the three possible outcomes for a pair of linear equations and to match the algebraic result with the geometry of the graphs. You might post this information:

Unique solution (independent system) ↔ Intersecting lines

No solution (inconsistent system) ↔ Parallel lines

Infinitely many solutions (dependent system) ↔ The same line

3. *"How Many of Each Kind?" Revisited*

Tell students they are now ready to return to the central unit problem, *How Many of Each Kind?* Have them look at the new activity on this problem, *"How Many of Each Kind?" Revisited*. Point out that the problem itself is unchanged, and have them look over the part called "Your Assignment" carefully. Emphasize that you want a good written presentation of their reasoning.

We suggest that you give groups the rest of today's class and all of tomorrow to produce these reports. You can have presentations on Day 21.

You may want to remind students that they did considerable work earlier in the unit toward solving this problem (see Days 1 through 7). Tell them that they can use that earlier work as an aid, but that their reports should explain everything from scratch.

Note: You may want to tell students that their write-ups of this activity should be included in their portfolios.

Homework 19: A Charity Rock

Part I of this homework is algebra practice, and Part II is a straightforward problem involving two equations in two unknowns.

"How Many of Each Kind?" Revisited

As you saw in *How Many of Each Kind?* Abby and Bing Woo have a small bakery that makes two kinds of cookies—plain cookies and cookies with icing. Here is a summary of the key information about the Woos' situation.

Summary of the Situation

Facts about making the cookies:

- Each dozen plain cookies requires 1 pound of cookie dough.
- Each dozen iced cookies requires 0.7 pounds of cookie dough and 0.4 pounds of icing.
- Each dozen plain cookies requires 0.1 hours of preparation time.
- Each dozen iced cookies requires 0.15 hours of preparation time.

Constraints:

- The Woos have 110 pounds of cookie dough and 32 pounds of icing.
- The Woos have room to bake a total of 140 dozen cookies.
- The Woos together have 15 hours available for cookie preparation.

Costs, prices, and sales:

- Plain cookies cost $4.50 a dozen to make and sell for $6.00 a dozen.
- Iced cookies cost $5.00 a dozen to make and sell for $7.00 a dozen.
- No matter how many of each kind they make, the Woos will be able to sell them all.

The Big Question is:

How many dozens of each kind of cookie should Abby and Bing make so that their profit is as high as possible?

Continued on next page

Your Assignment

Imagine that your group is a business consulting team, and the Woos have come to you for help. Of course, you want to give them the right answer. But you also want to explain to them clearly how you know that you have the best possible answer so that they will consult your group in the future.

You may want to review what you already know from earlier work on this problem. Look at your notes and earlier assignments. Then write a report for the Woos. Your report should cover these items.

- An answer to the Woos' dilemma, including a summary of how much cookie dough, icing, and preparation time they will use, and how many dozen cookies they will make altogether
- An explanation for the Woos that will convince them that your answer gives them the most profit
- Any graphs, charts, equations, or diagrams that are needed as part of your explanation

You should write your report based on the assumption that the Woos do not know the techniques you have learned in this unit about solving this type of problem.

Homework 19 A Charity Rock

Part I: Solving Systems

Solve each pair of equations.

1. $5x + 2y = 11$ and $x + y = 4$
2. $2p + 5q = 15$ and $6p + 15q = -29$
3. $3a + b = 4$ and $6a + 2b = 8$

Part II: Rocking Pebbles

At concerts given by the group Rocking Pebbles, some of the tickets sold are for reserved seats and the rest are general admission.

For a recent series of two weekend concerts, the Pebbles pledged to give their favorite charity an amount equal to half of what was paid for general-admission tickets. After the concerts, the charity called the Pebbles' manager to find out how much money the charity would be getting.

The manager looked up the records. She found that for the first night, 230 reserved-seat tickets and 835 general-admission tickets were sold. For the second night, 250 reserved-seat tickets and 980 general-admission tickets were sold.

The manager saw that the total amount of money collected for tickets was $23,600 for the first night and $27,100 for the second night, but she didn't know the prices for the two different kinds of tickets. (The prices were the same at both concerts.)

Figure out what the two ticket prices were, and use that information to tell the manager how much the Pebbles will give to the charity. As part of your work on this problem, set up a pair of linear equations. Then solve this system in whatever way you like, such as using algebra, graphs, or guess-and-check.

DAY 20 Finishing Cookies

Students finish the unit problem and prepare for tomorrow's presentations.

Mathematical Topics

- Continuing to work with linear systems, including inconsistent and dependent systems
- Using two equations in two unknowns to represent a situation
- Writing a complete analysis and explanation of a complex linear programming problem

Outline of the Day

In Class

1. Discuss *Homework 19: A Charity Rock*
   - Use Part I to review the concepts of inconsistent and dependent equations
   - Use Part II to focus on defining variables carefully and developing equations to describe the situation
2. Continue work on *"How Many of Each Kind?" Revisited* (from Day 19)
   - Students need to prepare presentations for Day 21

At Home

Homework 20: Back on the Trail

Discuss With Your Colleagues

Isn't Skill Practice Un-IMP-like?

Over the past several days, students have done a lot of work on algebraic skills without much context—at least, a lot for an IMP unit. Is this appropriate?

1. Discussion of *Homework 19: A Charity Rock*

- *Part I: Solving Systems*

Ask the students to come to a consensus in their groups about the solutions to Part I of the homework. As students are working, circulate among the groups to get a feel for how much time you need to spend going over this homework.

Even if students don't seem to have trouble, it's probably worth having presentations on both Question 2 (an inconsistent system) and Question 3 (a dependent system).

- *Part II: Rocking Pebbles*

Have one or two club card students present their solutions to Part II. You may need to remind students that finding the two ticket prices does not mean they are finished with the problem. They also need to find half the proceeds from the general-admission tickets.

Make sure that presenters define their variables clearly—for example, "x = *the price of* reserved-seat tickets" rather than "x = reserved seats."

Once they have the equations, students may use either the process that they developed in *Get the Point* or more of an estimation/graphing approach. Both are worthwhile.

2. Continued Work on *"How Many of Each Kind?" Revisited*

For the rest of today's class time, students should work on their "cookies" reports. The solution will be discussed tomorrow. Choose one or two groups to prepare presentations.

Homework 20: Back on the Trail

The problems in tonight's homework are somewhat traditional two-equation/two-unknown problems. The first is easier, because one of the equations expresses one of the variables directly in terms of the other. One of the main purposes of this homework is to give students some more models of what such a problem might look like, because their homework tomorrow night will be to invent one.

Homework 20 Back on the Trail

The two problems here are similar to ones that you saw in the *Overland Trail* unit. But now you know more about writing and solving equations, so these should be easier than when you first saw them.

For each of these two problems, write a pair of equations using two variables, and then solve the equations to answer the question.

Part I: Fair Share on Chores

A family with three boys and two girls needs to split up the chore of watching the animals. Altogether, the animals need to be watched for 10 hours.

If the length of each boy's shift is an hour more than the length of each girl's shift, how long is each type of shift? (Remember that the family considered this fair in light of other chores the boys and girls had to do.)

Part II: Water Rationing

The Stevens family contains three adults and five children. The Muster family contains two adults and four children. In a typical day on the trail, the Stevenses use about 15 gallons of water (for drinking, washing, and so on), while the Musters use about 11 gallons per day.

Assuming that each adult uses about the same amount and that each child uses about the same amount, how much does each use in a typical day?

DAY 21 *Cookie Presentations*

Students present solutions to the unit problem and begin a new problem.

Mathematical Topics

- Presenting the solution to the unit problem
- Solving traditional two-equation/two-unknown word problems

Outline of the Day

In Class

1. Presentations of *"How Many of Each Kind?" Revisited*
   - Focus the discussion on use of the family of parallel profit lines
2. Discuss *Homework 20: Back on the Trail*
   - Discussion can be brief
3. Compare two types of problems
   - Help students distinguish linear programming problems from two-equation/ two-unknown problems
   - Use *Homework 12: Rock 'n' Rap* to illustrate the connection between the two types
4. Introduce *Homework 21: Inventing Problems*
5. *Big State U*
   - Students work on one more new linear programming problem to deepen their understanding before devising their own problems
   - The activity will be discussed on Day 22

At Home

Homework 21: Inventing Problems

Note: We recommend that you delay discussion of last night's *Homework 20: Back on the Trail* until after presentations of *"How Many of Each Kind?" Revisited* so that the homework discussion can lead into the discussion of two-variable/two-equation problems and tonight's assignment, *Homework 21: Inventing Problems*.

1. Presentations of *"How Many of Each Kind?" Revisited*

Have one or two groups present their analysis of *"How Many of Each Kind?" Revisited*.

The graph shown here is the same as the graph of the feasible region given on Day 7, except that one of the profit lines has been added (represented by the dashed line). This line is the graph of the equation $1.5P + 2I = 150$ and shows those combinations of dozens of plain cookies and dozens of iced cookies that give a total profit of $150.

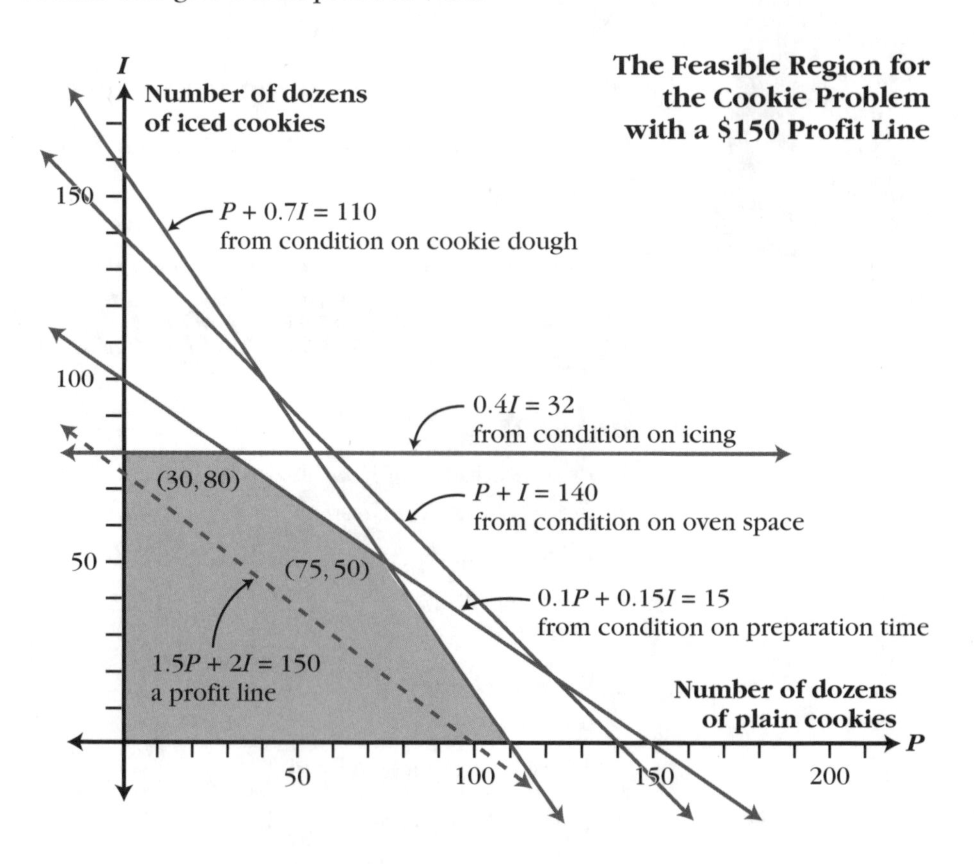

Some questions you can ask:
"What does each line represent?"
"How do you determine the feasible region?"
"What expression describes profit?"
"How do you know that (75, 50) is the best choice?"

At this stage, students should definitely include the coordinates of the key points shown—(30, 80) and (75, 50). They should be able to explain what each of the lines represents and how the feasible region is determined. For example, they should be able to explain why we take the area *below* rather than *above* a certain line. Finally, they should be able to identify the expression $1.5P + 2I$ as describing the profit and to explain why the best choice for the Woos is to make 75 dozen plain cookies and 50 dozen iced cookies.

Students will probably use the "family of parallel lines" reasoning. That is, the points that give any particular profit are on a straight line, and as the profit increases, the line is replaced by a parallel one above and to the right. The last of these lines to intersect the region is the one through the point (75, 50).

You can ask students how they can be sure where the "parallel family" leaves the region. For example, how do they know it isn't at the point (30, 80)? They might respond that the answer must be at either (30, 80) or (75, 50), because these are corners, and then simply evaluate the profit expression $1.5P + 2I$ at both points to see which is better. The profit at (30, 80) is $205.00, while the profit at (75, 50) is $212.50.

2. Discussion of *Homework 20: Back on the Trail*

Have one or two students present solutions to the problems. If neither presenter took a standard two-equation/two-unknown approach, ask for a volunteer to do the problem that way, making sure that the variables are clearly defined.

In the first problem, using b for the number of hours in a boy's shift and g for the number of hours in a girl's shift, the equations will probably be

$$b = g + 1$$
$$3b + 2g = 10$$

The solution to this system is $b = 2.4$, $g = 1.4$, so each boy's shift should be 2.4 hours and each girl's shift should be 1.4 hours.

For the second problem, using a and c for the amount of water used by an adult and by a child respectively, the likely equations are

$$3a + 5c = 15$$
$$2a + 4c = 11$$

The solution is $a = 2.5$, $c = 1.5$. That is, each adult uses 2.5 gallons of water, and each child uses 1.5 gallons.

3. Two Types of Problems

"How do the homework problems compare with the cookie problem?"

"What are some other problems you've done that are similar to each type?"

Ask the class how the problems from last night's homework compare with the central unit problem, *How Many of Each Kind?* It may help to ask for other examples similar to each of these. Here are some examples you can bring up if students are at a loss.

- Problems similar to *Homework 20: Back on the Trail*
  - Part II of *Homework 19: A Charity Rock*
  - Question 1 of *Homework 17: A Reflection on Money*
  - *Homework 14: Going Out for Lunch*
- Problems similar to *How Many of Each Kind?*
  - *Homework 12: Rock 'n' Rap*
  - *Profitable Pictures* (from Days 8 and 9) *and Hassan's a Hit!* (from Day 10)
  - *Homework 10: You Are What You Eat*

The problems in both lists are all similar in that they involve linear equations and require finding points of intersection of lines (or, equivalently, the solution to a system of two linear equations). Help students to see, though, that there are really two different types of problems in this unit:

- Two-equation/two-unknown problems (like the homework)
- Maximizing (or minimizing) problems (like the unit problem)

Ask them to describe these two types as clearly as they can.

In the first type, the entire situation is described by a pair of linear equations, and the task is simply to find the common solution, which is the point of intersection of their graphs.

The second type actually begins with several linear *inequalities*, which are used to define a *feasible region*. The task is to maximize (or minimize) some linear *expression* within that region. Students used a geometric argument, involving a family of parallel lines, to find out which intersection point they were looking for, and only at that stage did they solve a pair of linear equations.

• *Looking back at Homework 12: Rock 'n' Rap*

It might be helpful for students to refer back to one or two examples of the second type of problem and have them interpret, in the context of the problem, what "solving the equations" meant.

For example, in *Homework 12: Rock 'n' Rap*, there was a profit function, $20{,}000x + 30{,}000y$, as well as these constraint inequalities:

$$y \leq x$$

$$15{,}000x + 12{,}000y \leq 150{,}000$$

$$18x + 25y \geq 175$$

$$x \geq 0$$

$$y \geq 0$$

Graphing these inequalities led to a feasible region whose boundaries were formed from the graphs of the corresponding equations.

The main task in the problem was figuring out, based on the graphs, which two of these five equations should be used to find the point that maximized profit. Once students knew which point they wanted, they found the common solution to the appropriate two equations. Thus, the last step in solving the problem was finding the common solution to the equations $y = x$ and $15{,}000x + 12{,}000y = 150{,}000$.

Tell students that this type of problem is called a **linear programming problem** and that they will look more at such problems in tomorrow night's homework.

- *Making a similar two-equation/two-unknown problem*

Ask students to create a question based on the situation in *Homework 12: Rock 'n' Rap*, but in which they could simply start with the pair of equations $y = x$ and $15{,}000x + 12{,}000y = 150{,}000$. They should come up with something like this:

> Hits on a Shoestring wants to make an equal number of rock and rap CDs. Each rock CD costs $15,000 to produce, and each rap CD costs $12,000. The company spends a total of $150,000. How many of each kind of CD does it make?

Students should see that this problem has nothing to do with profit, even though finding the point of intersection of the two lines was a step in figuring out how Hits on a Shoestring could maximize its profit.

"Why is this called a two-equation/two-unknown word problem?"

Describe the word problem just created as a **two-equation/two-unknown** word problem. Be sure students see why it is called that.

4. Introduction to *Homework 21: Inventing Problems*

Have students look at tonight's assignment, *Homework 21: Inventing Problems*. Point out that the problem they just made up for the equations $y = x$ and $15{,}000x + 12{,}000y = 150{,}000$ is an example of the type of problem they are to create for homework.

You might mention, in connection with Question 3, that "professional problem writers" (people who write traditional textbooks or IMP units, for example) often come up with the general idea for a problem, write possible equations, and then go back and adjust the numbers in the problem so that the equations and solutions are reasonable.

You might mention as well that tomorrow students will begin an assignment, working in groups, in which they create their own linear programming problems.

5. *Big State U*

Big State U is the last of the linear programming problems in the unit. This problem will give students an opportunity to synthesize their ideas about how to solve complex problems of this type before they write one of their own.

Homework 21: Inventing Problems

Tonight's assignment is partly a lead-in to the more complex task, *Producing Programming Problems,* touched on tomorrow and introduced formally on Day 23.

Big State U

The Admissions Office at Big State University needs to decide how many in-state students and how many out-of-state students to admit to the next class. Like many universities, Big State U has limited resources, and budget considerations have to play a part in admissions policy.

Here are the constraints on the Admissions Office decision.

- The college president wants this class to contribute a total of at least \$2,500,000 to the school after it graduates. In the past, Big State U has gotten an average of \$8,000 in contributions from each in-state student admitted and an average of \$2,000 from each out-of-state student admitted.
- The faculty at the college wants entering students with good grade-point averages. Grades of in-state students average less than grades of out-of-state students. Therefore, the faculty is urging the school to admit at least as many out-of-state students as in-state students.
- The housing office is not able to spend more than \$85,000 to cover costs such as meals and utilities for students in dormitories during vacation periods. Because out-of-state students are more likely to stay on campus during vacations, the housing office needs to take these differences into account. In-state students will cost the office an average of \$100 each for vacation-time expenses, while out-of-state students will cost an average of \$200 each.

The college treasurer needs to minimize educational costs. Because students take different courses, it costs an average of \$7,200 a year to teach an in-state student and an average of \$6,000 a year to teach an out-of-state student.

Your job is to recommend how many students from each category should be admitted to Big State U. You need to minimize educational costs, as the treasurer requires, within the constraints set by the college president, the faculty, and the housing office.

Your write-up should include a proof that your solution is the best possible within the constraints. Show any graphs that seem helpful, and explain your reasoning carefully.

Adapted from *An Introduction to Mathematical Models in the Social and Life Sciences,* by Michael Olinick, Addison-Wesley, 1978, p. 169.

Homework 21 Inventing Problems

You have now seen several problems that you could solve by defining variables and then setting up and solving a pair of linear equations. Examples include *Homework 14: Going Out for Lunch, Homework 19: A Charity Rock,* and *Homework 20: Back on the Trail.*

In this assignment, you get to make up your own problem.

1. Make up a problem that you think can be solved with two equations and two unknowns.
2. Solve the problem and write up your solution. (*Note:* As you work on the problem, you may find that you want to change it in some way to improve it.)
3. Write out your problem (without solution) on a separate sheet of paper. Put your name on this sheet. (Tomorrow, students will share and work on one another's problems.)

Cookies

Creating Problems

This page in the student book introduces Days 22 through 27.

Over the course of this unit, you have solved a variety of problems involving linear equations, linear inequalities, and graphs. One way to get more insight into such problems is to create one of your own. In *Homework 21: Inventing Problems,* you created a two-equation/two-unknown problem. In the final days of the unit, you will create a linear programming problem.

Ryan Jones, Keith Landrum, and Mandy Ledford prepare the graph for the linear programming problem they created.

DAY 22 Sharing Problems and Big State U

Students share the problems they invented and discuss a linear programming problem.

Mathematical Topics

- Inventing problems that use linear equations
- Solving and presenting solutions to a linear programming problem

Outline of the Day

In Class

1. Discuss *Homework 21: Inventing Problems*
   - Have each group present its best problem
2. Discuss *Big State U* (from Day 21)
   - Continue to focus on the use of the family of parallel lines
3. Introduce *Homework 22: Ideas for Linear Programming Problems*
   - Review the distinction between linear programming problems and two-equation/two-unknown problems

At Home

Homework 22: Ideas for Linear Programming Problems

1. Discussion of *Homework 21: Inventing Problems*

Have students work in groups on one anothers' problems. Each group should pick its best problem to present to the class. Give each group an overhead transparency and pens. We recommend that you tell students that you will grade the groups on the presentations and that the presenter will be picked at random from the group.

When groups are ready, choose a suit at random and have that student from each group present the group's best problem, with solution.

2. Discussion of *Big State U*

Let students conclude their work on *Big State U.* Choose one or two groups that did not make presentations on the unit problem to make presentations on this activity.

- ***The basics of the solution***

Letting x represent the number of in-state students to be admitted and y represent the number of out-of-state students, the constraints are:

$8000x + 2000y \geq 2{,}500{,}000$	(for the contributions the president wants)
$y \geq x$	(to satisfy the faculty's concern about grades)
$100x + 200y \leq 85{,}000$	(to fit the housing office's budget)
$x \geq 0,\ y \geq 0$	(because the values can't be negative)

The expression that needs to be minimized, for the sake of the treasurer, is $7200x + 6000y$.

The diagram on the facing page shows the graphs of the equations that go with the constraints, as well as the feasible region (the small darkly-shaded area). (If the president had required a little more in contributions, or if the housing office had had a little less money, the situation would have been impossible.)

The dashed line is from the family of parallel lines that represent different cost conditions. This particular line shows the possible combinations of enrollments that give a cost of $1,440,000; that is, the dashed line is the graph of the equation $7200x + 6000y = 1{,}440{,}000$. The diagram shows that the minimum "cost line" to intersect the feasible region is the one that goes through the intersection of the lines $y = x$ and $8000x + 2000y = 2{,}500{,}000$. (Details on the reasoning follow on the next page.)

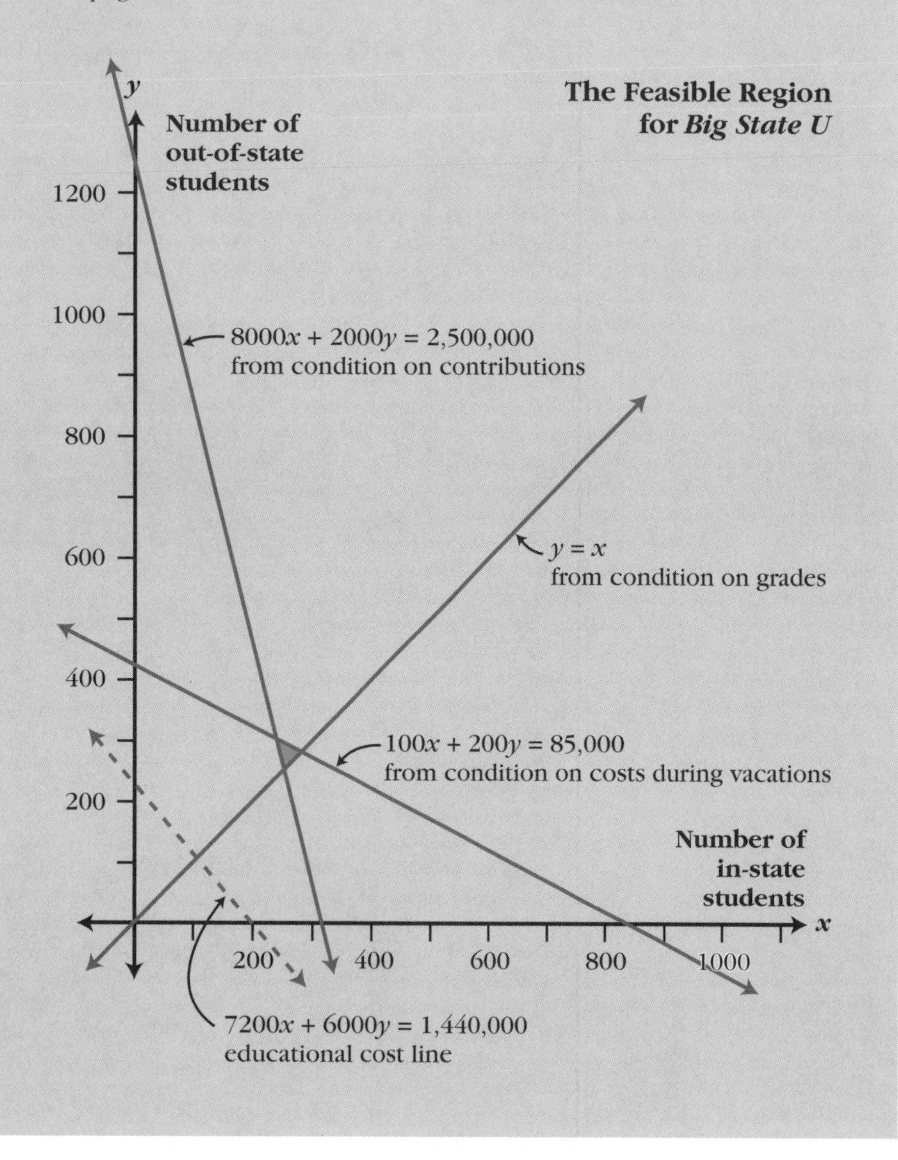

- ***A close-up of the feasible region***

This diagram shows the feasible region in more detail and gives the coordinates of the points of intersection. The point that minimizes educational costs is (250, 250). Thus, the school should admit 250 in-state students and 250 out-of-state students.

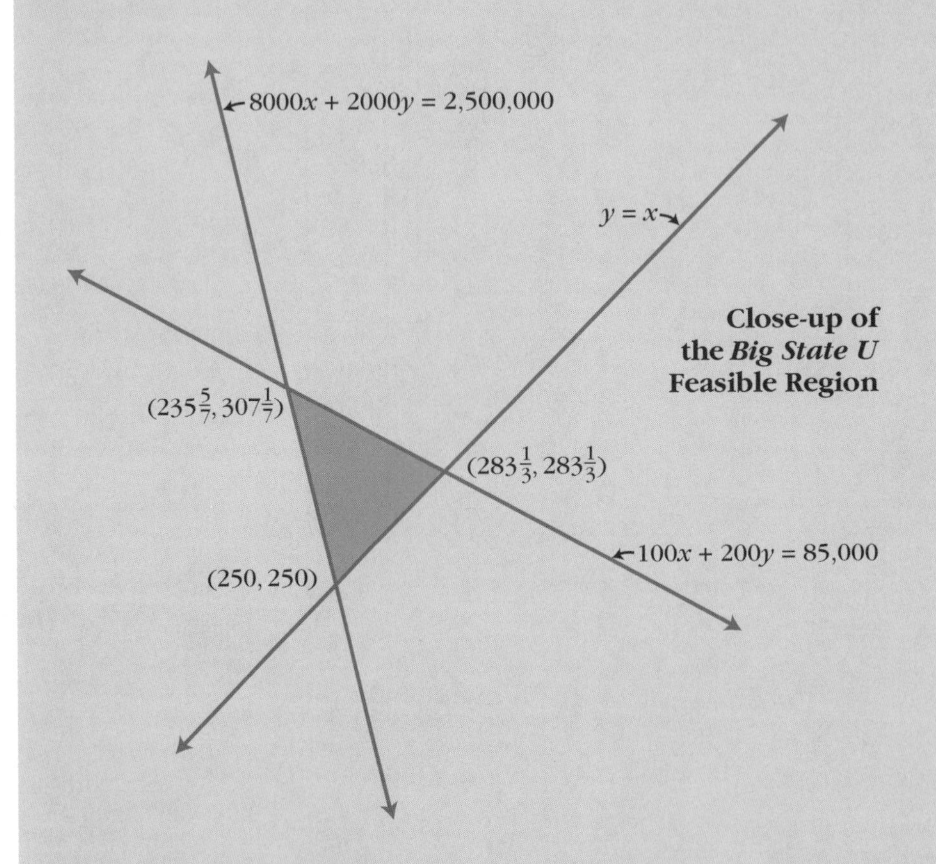

- ***The details of the reasoning***

In presenting the problem, students should review how they sketched the graph, including how they knew which point minimized the cost and how they got its coordinates.

In explaining how they identified the point, students should use the "family of parallel lines" idea, in which the combinations of in-state students and out-of-state students that give any particular cost form a straight line, which "slides" when the cost is changed.

Students need to choose the "minimum" of this family of parallel lines. If they sketch one of these lines (for example, $7200x + 6000y = 1{,}440{,}000$), they will see the general direction of the lines in the family. (This is the dashed line in the first diagram.)

Students should see that the cost goes up as the line moves up and to the right, and so the minimum is at the point (250, 250).

3. Introduction to *Homework 22: Ideas for Linear Programming Problems*

You can introduce tonight's homework by reminding students that yesterday they distinguished between two types of problems:

- Two-equation/two-unknown problems
- Maximizing (or minimizing) problems, called *linear programming problems*

Point out that in last night's homework, students invented problems of the first type. Tell them that in the final activity in the unit, *Producing Programming Problems,* each group will make up a linear programming problem and present its solution to the class.

Tonight's assignment is intended to give students ideas for those problems. It explains in more detail what a linear programming problem is and asks students to examine some previous problems of this type to get a clearer sense of what they have in common and how they work.

Mark Hansen, Jennifer Rodriguez, Karla Viramontes, and Robin Lefevre work on their graph of the feasible region for "Big State U."

Homework 22 Ideas for Linear Programming Problems

In tomorrow's activity, *Producing Programming Problems,* your group will be creating a linear programming problem to solve and present to the class.

In any linear programming problem, there are variables, constraints, and something to minimize or maximize. For instance, the central problem of this unit (*How Many of Each Kind?*) is a linear programming problem. The variables represented the number of dozen plain cookies and the number of dozen iced cookies. The constraints were the Woos' available preparation time and oven space and the amounts of cookie mix and icing mix. The goal was to maximize the profit.

Big State U is another linear programming problem. In that problem, the task was to minimize the educational costs, using variables representing the number of in-state students and the number of out-of-state students to be admitted. The constraints involved contributions to the university, grade-point averages, and housing costs.

1. Reread *Homework 10: You Are What You Eat.*
   a. What did the variables in that problem represent?
   b. What were the constraints?
   c. What needed to be maximized or minimized?

Continued on next page

2. Reread the activity *Profitable Pictures.*
   a. What did the variables in that problem represent?
   b. What were the constraints?
   c. What needed to be maximized or minimized?
3. Reread *Homework 12: Rock 'n' Rap.*
   a. What did the variables in that problem represent?
   b. What were the constraints?
   c. What needed to be maximized or minimized?
4. Create a situation in which you might be interested in maximizing or minimizing something. Describe the situation, and state what you would maximize or minimize.
5. Choose two variables to go with your situation from Question 4, and give two or three constraints using those variables that might apply.

DAY 23 Creating a Problem

Groups begin to create their own linear programming problems.

Mathematical Topics

- Understanding the elements of linear programming
- Identifying situations suitable for the use of linear programming techniques

Outline of the Day

In Class

1. Select presenters for tomorrow's discussion of *POW 13: Shuttling Around*
2. Discuss *Homework 22: Ideas for Linear Programming Problems*
   - Emphasize the linearity of constraint and profit expressions
   - Get (and perhaps post) a list of topic ideas
3. *Producing Programming Problems*
   - Groups create their own linear programming problems
   - Discuss grading criteria
   - The activity will continue on Days 24 and 25, with presentations on Days 26 and 27

At Home

Homework 23: Beginning Portfolio Selection

Discuss With Your Colleagues

Is the Time Spent on the Project Worthwhile?

Students will be spending nearly a week preparing and presenting work on *Producing Programming Problems.* What might your students get out of this project? Do the results justify the time spent?

1. POW Presentation Preparation

Presentations of *POW 13: Shuttling Around* are scheduled for tomorrow. Choose three students to make POW presentations, and give them overhead transparencies and pens to take home to use in their preparations.

You may want to suggest that they bring in objects of two different sizes that could be used on the overhead projector to represent the two types of markers.

2. Discussion of *Homework 22: Ideas for Linear Programming Problems*

Have different heart card students present their answers for Questions 1, 2, and 3.

"Why are these called linear programming problems?"

Ask students why the word "linear" is used in the phrase "linear programming" to describe these problems. They should identify two distinct components of the problem that are linear:

- The constraints
- The expression being maximized or minimized

Comment: The "programming" part of the term "linear programming" seems to refer to the fact that there is a systematic method—a program in the nontechnical, noncomputer sense—for solving problems of this type.

Whether or not you discuss the optional topic below, make clear to students that for their own good, they should not use nonlinear constraints or expressions in their problems, because these may vastly complicate the process of solving the problems.

- ***(Optional) Why is linearity important?***

You can discuss why linearity is important in these problems. There are two main reasons.

- Because the constraints are linear, the feasible region is a polygon, and its vertices can be found by solving a pair of linear equations.
- Because the expression to be maximized (or minimized) is linear, the points that give this expression a given value lie on a straight line, and we can then look at a family of parallel lines.

You can mention that there are certainly problems or situations in which the expression being maximized or minimized is nonlinear or in which the constraints are nonlinear, but both the algebra and the geometry are much more complicated when this happens.

• *Questions 4 and 5*

Ask for volunteers to share their ideas on Questions 4 and 5. You can list them on chart paper and post them so that students can refer to these ideas when they begin work on *Producing Programming Problems.*

It is important to emphasize linearity in this discussion. With each example students give, focus on whether the "thing being maximized (or minimized)" depends *linearly* on the variables they have chosen. If not, try to elicit a way to change the suggestion so that it does do so.

Also look at whether the constraints are linear conditions on the variables selected.

• *For teachers: More on linear programming*

The linear programming problems that students have been doing are the most elementary type, primarily because these problems involve only two variables and a small number of constraints.

For more complex problems, the task of finding the right vertex is vastly more complicated. In many applications, the problems are solvable only through the use of supercomputers and the application of techniques that help search for the right vertex.

The best known such technique, called the *simplex method,* was invented about 40 years ago. Credit for inventing this technique is generally given to George Dantzig, a professor of operations research at Stanford University.

The simplex method and other techniques are widely used today to solve linear programming problems in many industries.

3. *Producing Programming Problems*

Once the discussion of *Homework 22: Ideas for Linear Programming Problems* is completed, have students turn to *Producing Programming Problems*. Students can use the rest of today to begin work on this activity.

Presentations on *Producing Programming Problems* are scheduled to begin on Day 26. Except for tomorrow's POW discussion and the Day 25 discussion of *Homework 24: Just for Curiosity's Sake*, students have all of

Day 24 and Day 25 to work on the activity. In tonight's *Homework 23: Beginning Portfolio Selection,* they are asked to reflect on the process of solving linear programming problems. You need not take discussion time tomorrow for this, but it should be helpful in students' work on *Producing Programming Problems.*

You may want to adjust the time allotted for *Producing Programming Problems* based on your own students' needs.

Tell students that this assignment is going to be part of their unit portfolios, so it is essential that each student do his or her own write-up. Their assignment in *Homework 25: "Producing Programming Problems" Write-Up* is to complete their write-ups.

- *Reviewing student proposals for linear programming problems*

You might set up a process for reviewing students' problem ideas before they get very far into them. You may want to examine them in terms of both mathematical appropriateness and appropriateness of the subject matter.

- *Peer grading and grading criteria*

We recommend that you have groups grade the presentations on this activity, including their own, and we give some ideas for the mechanics of this on Day 26. You should discuss with students how you will use their grades of themselves and one another.

Whether or not you do peer grading, you should discuss grading criteria with the class, perhaps letting them develop their own criteria. If they need ideas, here are some areas that you can suggest:

- Creativity
- Mathematical elegance
- Clarity of presentation
- Participation of all group members

Post the grading criteria so students can refer to them while they work on the problems and during the presentations.

- *Optional: Videotape the presentations*

Many teachers who have videotaped these project presentations report that this has given the activity added importance for students.

Homework 23: Beginning Portfolio Selection

Tonight's homework assignment gets students started compiling their portfolios for this unit, asking them to summarize what they know about solving linear programming problems. This process should help students develop their own linear programming problems.

Producing Programming Problems

Your group is to make up a linear programming problem. Here are the key ingredients you need to have in your problem.

- Two variables
- Something to be maximized or minimized that is a *linear function* of those variables
- Three or four *linear* constraints

Once you have written your problem, you must solve it.

Then you should put together an interesting 5-to 10-minute presentation. This presentation should do three things.

- Explain the problem
- Provide a solution to the problem
- Prove that there is no better solution

Homework 23 Beginning Portfolio Selection

The main problem for this unit, *How Many of Each Kind?* is an example of a **linear programming** problem. You have seen several such problems, including *Profitable Pictures*, *Homework 10: You Are What You Eat*, *Homework 12: Rock 'n' Rap*, and *Big State U.*

1. Describe the steps you must go through to solve such a problem.
2. Pick three activities from the unit that helped you to understand particular steps of this process, and explain how each activity helped you. (You do not need to restrict yourself to the activities listed above.)

   Note: Selecting these activities and writing the accompanying explanations are the first steps toward compiling your portfolio for this unit.

DAY 24

POW 13 Presentations

Students present POW 13, and groups continue creating their linear programming problems.

Mathematical Topics

- Finding a function to describe patterns from a puzzle
- Developing formulas from examples
- Developing a context for applying linear programming techniques

Outline of the Day

In Class

1. Discuss *Homework 23: Beginning Portfolio Selection*
   - Have a few students share their ideas
2. Presentations of *POW 13: Shuttling Around*
   - Ask about a general formula for the number of moves
   - If students have a formula, ask for a proof
3. Continue work on *Producing Programming Problems* (from Day 23)

At Home

Homework 24: Just for Curiosity's Sake

1. Discussion of *Homework 23: Beginning Portfolio Selection*

You can have a couple of volunteers read their descriptions of the overall process for solving linear programming problems. You may also want to discuss what makes a good description. Finally, you might have some students share their choice of activities in Question 2 of the assignment and their explanations of how these activities helped them understand the ideas.

2. Presentations of *POW 13: Shuttling Around*

Have the students selected yesterday make their presentations, and let other students add further ideas.

"How might you get a general formula for the number of moves?"

If no one presents a general formula for the number of moves required, you can ask students how they might develop such a formula. If needed, suggest compiling all the data presented into an In-Out table.

"How might you prove the general formula?"

This may lead some students to see a formula, but if not, you can leave it as an open problem. If students do develop a formula, ask if anyone has a way to explain or prove the formula. Again, if no one has ideas, you need not push for an answer.

Note: The supplemental problem *Shuttling Variations* is an excellent follow-up to this POW.

3. Continued Work on *Producing Programming Problems*

Students should use whatever time remains for work in groups developing their linear programming problems. *Reminder:* They will have all of tomorrow's classtime as well, except for the time spent discussing the homework.

Homework 24: Just for Curiosity's Sake

The problem in Part II of this homework has no solution, because the two equations are inconsistent. But don't warn students. This will be brought up in tomorrow's homework discussion.

Homework 24 Just for Curiosity's Sake

Part I: Solving Equations

Solve each pair of equations.

1. $3s + t = 13$ and $2s - 4t = 18$
2. $6(a + 2) - b = 31$ and $5a - 2(b - 3) = 23$
3. $z - w = 6$ and $5z + 3w = 10$

Part II: Rocking Pebbles

The Rocking Pebbles just finished another two-night series of concerts in a different town. This time, none of the shows was for charity, because the producer in that town thought doing so might set a bad precedent. The Pebbles' manager wonders how the producer priced the tickets. (The prices may have been different from those in the previous town but they were the same both nights.)

The producer said there were 200 reserved-seat tickets and 800 general-admission tickets sold the first night, and that the total money taken in was $20,000. He said that on the second night, they sold 250 reserved-seat tickets and 1000 general-admission tickets, and that the total money taken in that night was $23,000.

To satisfy the curiosity of the Pebbles' manager about the producer's pricing policy, find out the cost of reserved-seat tickets and the cost of general-admission tickets. Include a pair of equations that will describe the problem.

Producing Programming Problems, Concluded

Groups finish creating their linear programming problems.

Mathematical Topics

- Developing a context for applying linear programming techniques

Outline of the Day

In Class

1. Discuss *Homework 24: Just for Curiosity's Sake*
   - Use Part II to review the idea of inconsistent equations
2. Conclude work on *Producing Programming Problems* (from Day 23)
   - Groups should prepare their presentations for Days 26 and 27

At Home

Homework 25: "Producing Programming Problems" Write-up

1. Discussion of *Homework 24: Just for Curiosity's Sake*

- *Part I: Solving Equations*

Ask students to check their answers in their groups for the systems in Part I of the homework. Unless there are difficulties, you don't need to spend time on presentations of them.

The answers are

- Question 1: $s = 5$ and $t = -2$
- Question 2: $a = 3$ and $b = -1$
- Question 3: $w = -2.5$ and $z = 3.5$

- *Part II: Rocking Pebbles*

Students should see that there is no pricing policy that will give the results stated. Nevertheless, they should be able to write a pair of equations that describe the situation. For example, using R for the cost of a reserved seat ticket and G for the cost of a general-admission ticket, their equations would be

$$200R + 800G = 20{,}000 \quad \text{(for the first night)}$$

$$250R + 1000G = 23{,}000 \quad \text{(for the second night)}$$

Get students to articulate that these two equations have graphs that are parallel lines, which means there is no common solution.

"How could you use equivalent equations to see that these equations are clearly inconsistent?"

Ask students how they might show algebraically that these two equations are "obviously" inconsistent, perhaps using equivalent equations. One approach is to divide the first by 200 and the second by 250, giving $R + 4G = 100$ and $R + 4G = 92$, respectively. Another is to see that the number of tickets sold of each type went up exactly 25 percent on the second night but the money taken in did not.

- *Optional: Extension of Part II*

If time allows, you can pose this follow-up question after the inconsistency of the equations has been pointed out.

> When the inconsistency in the receipts was pointed out, the bookkeeper checked the records. Sure enough, the amount taken in on the second night was actually \$25,000, not \$23,000. What does that tell you about the prices of the two types of tickets?

Let students work on this in their groups for a few minutes. They should see that they still can't determine the prices, because the new second equation,

$$250R + 1000G = 25{,}000$$

has the same graph as the first equation, $200R + 800G = 20{,}000$. (Both equations simplify to $R + 4G = 100$.)

Students should realize that *any point* on that graph will satisfy the equation (with perhaps the condition that $R > G$, because reserved-seat tickets should cost more than general-admission tickets). But they should also see that there is not enough information to decide exactly which point represents the actual prices. You can use this as an occasion to review the term *dependent equations.*

2. Conclusion of Work on *Producing Programming Problems*

Students should spend the rest of the class time completing their linear programming problems and preparing for their presentations. You may want to have a short discussion on what would make an excellent presentation. Such a discussion could be preceded by some focused free-writing.

Presentations are scheduled for the next two days. You may want to decide today which groups will do presentations on which day, or you may prefer to choose groups at random tomorrow.

Homework 25: "Producing Programming Problems" Write-up

Students' only homework tonight is to complete their write-ups for their work on *Producing Programming Problems.*

Tell students that write-ups for *Producing Programming Problems* will be part of their portfolios for this unit.

If you have told students which day their presentation will be, you may want to suggest that groups presenting on Day 27 do *Homework 26: Continued Portfolio Selection* tonight and do this assignment tomorrow night.

Homework 25 "Producing Programming Problems" Write-up

Your group should now have completed development of its own linear programming problem and prepared its presentation for the class. Your homework tonight is to complete your own write-up for the assignment. As with your group's presentation, this should include three things.

- A statement of the problem
- The solution
- A proof that the solution is the best possible

"Producing Programming Problems" Presentations

The first groups make presentations on the problems they created.

Mathematical Topics

- Giving and grading oral presentations

Outline of the Day

In Class

Presentations of *Producing Programming Problems*
- Groups work together to grade each presentation

At Home

Homework 26: Continued Portfolio Selection

Special Materials Needed

- (Optional) *Grading Programming Problem Presentations* (see Appendix B)

1. Presentations of *Producing Programming Problems*

Today and tomorrow are set aside for presentation of groups' work on *Producing Programming Problems.*

Before presentations begin, remind students of the criteria to be used for grading the presentation, which you probably have posted in the classroom. If students are grading one another's presentations, we suggest that you have them do this immediately after each presentation. All groups (including the presenting group) should decide on a letter grade and write a justification for that grade, based on the posted criteria. You may want to make copies of the grading form in Appendix B for students to use in this grading process.

Students should turn in their evaluations of the presentations to you. You can pass them on to the presenters after you have read them.

Be sure to allow the groups a reasonable amount of time to discuss and evaluate the quality of the presentation and to write their reports before moving on to the next presentation. You can use this time to assign each group your own grade.

You will probably find it best to collect the group grade reports on a given presentation before beginning the next one.

Homework 26: Continued Portfolio Selection

Tonight's assignment continues the work of preparing ***Cookies*** portfolios.

Sugey Ochoa puts the finishing touches on the chart for her group's linear programming problem.

Homework 26 Continued Portfolio Selection

In *Homework 23: Beginning Portfolio Selection,* you looked at one of the major themes of this unit—linear programming problems. In this assignment, you will look at another major theme—solving systems of linear equations with two variables.

1. Summarize what you learned about solving such systems, both in *Get the Point* and in work since that activity.
2. Choose two examples of problem situations that you could solve using a system of linear equations. Your examples can be from this unit or from an earlier unit. For each of the examples you give, explain how the algebraic representation of the problem using linear equations would help you solve the problem.

DAY 27

Finishing Presentations

Remaining groups make presentations on the problems they created.

Mathematical Topics

- Giving and grading oral presentations

Outline of the Day

In Class

Conclude presentations of *Producing Programming Problems*

At Home

Homework 27: "Cookies" Portfolio

Conclude Presentations

Finish the group presentations of *Producing Programming Problems* as described on Day 26.

Homework 27: "Cookies" Portfolio

For homework tonight, students will complete their unit portfolios. They will have done part of the selection process in *Homework 23: Beginning Portfolio Selection* and *Homework 26: Continued Portfolio Selection.* Their main task in this assignment is to write their cover letters.

Be sure that students bring back the portfolio tomorrow with the cover letter as the first item. They should also bring to class any other work they think will be of help on tomorrow's unit assessments. The remainder of their work can be kept at home.

Homework 27 *Cookies* Portfolio

Now that *Cookies* is completed, it is time to put together your portfolio for the unit. Compiling this portfolio has three parts.

- Writing a cover letter in which you summarize the unit
- Choosing papers to include from your work in this unit
- Discussing your personal growth during the unit

Cover Letter for "Cookies"

Look back over *Cookies* and describe the central problem of the unit and the main mathematical ideas. This description should give an overview of how the key ideas were developed and how they were used to solve the central problem.

Continued on next page

Selecting Papers from "Cookies"

Your portfolio for *Cookies* should contain

- *Homework 23: Beginning Portfolio Selection*
  Include the activities from the unit that you selected in *Homework 23: Beginning Portfolio Selection,* along with your written work about those activities that was part of the homework.
- *Homework 26: Continued Portfolio Selection*
- A Problem of the Week
  Select one of the three POWs you completed during this unit (*A Hat of a Different Color* or *Kick It!* or *Shuttling Around*).
- *Homework 17: A Reflection on Money*
- *Get the Point*
- *"How Many of Each Kind?" Revisited*
- *Producing Programming Problems*
  Include the statement and solution of the problem your group invented.

Personal Growth

Your cover letter for *Cookies* describes how the unit develops. As part of your portfolio, write about your own personal development during this unit. You may want to address this question:

> *How do you think you have grown so far in the area of making presentations?*

You should include here any other thoughts you might like to share with a reader of your portfolio.

DAY 28 Final Assessments

Students do the in-class assessment and begin work on the take-home assessment.

Outline of the Day

In Class

Introduce assessments

- Students do *In-Class Assessment for "Cookies"*
- Students begin *Take-Home Assessment for "Cookies"*

At Home

Students complete *Take-Home Assessment for "Cookies"*

Special Materials Needed

- *In-Class Assessment for "Cookies"*
- *Take-Home Assessment for "Cookies"*

End-of-Unit Assessments

Note: The in-class unit assessment is intentionally short so that time pressures will not affect students' performance. The IMP *Teaching Handbook* contains general information about the purpose of the end-of-unit assessments and ways to use them.

Tell students that today they will get two tests—one that they will finish in class and one that they can start in class and will be able to finish at home. The take-home part should be handed in tomorrow.

Tell students that they are allowed to use graphing calculators, notes from previous work, and so forth when they do the assessments. (They will have to do without graphing calculators when they complete the take-home portion at home unless they have their own.)

These assessments are provided separately in Appendix B for you to duplicate.

- *Clarifying Question 2a of the in-class assessment*

You should clarify, in terms of the specific calculators your students are using, what is being asked for on Question 2a of the in-class assessment. Students should be able to read off the values requested here using something like a "window" or "range" feature of their calculators. The values might be listed, for example, as Xmin, Xmax, Ymin, and Ymax.

Some teachers prefer to have students actually show them the graphing calculator with the graph displayed, as an alternative to having students provide the written response asked for in Question 2.

In-Class Assessment for "Cookies"

Part I: Graph It

Consider the following system of two equations:

$$5x + 7y = 200$$
$$9x - 3y = 85$$

1. Use your graphing calculator to estimate the solution to the system.
2. Set your viewing window so that the coordinate axes, the graphs of both equations, and the point of intersection of the two graphs are all visible.
   a. Give the values below to describe your viewing window.
      - Minimum value of x
      - Maximum value of x
      - Minimum value of y
      - Maximum value of y

   b. Make a sketch of what you see on your graphing calculator screen, and give the approximate coordinates of the solution.

Part II: Solve It

Use algebra to solve each system of equations. Show and explain your work clearly.

3. $4x + 3y = 5$
   $2x - 5y = 9$

4. $4x - 6y = 20$
   $6x - 9y = 24$

Take-Home Assessment for *Cookies*

Part I: What If . . . ?

The problems in this part of the assessment are variations on the unit problem, *How Many of Each Kind?* In each problem, you should find the combination of plain and iced cookies that maximizes the Woos' profit for the new situation, and then explain your answer. Consider these three variations from the original situation as *three separate problems.*

Each problem is accompanied by a graph that shows the feasible region of the original problem, with the shaded area representing that original feasible region.

Questions 1 and 2 show a profit line based on the original problem. Question 3 shows a profit line based on a different profit expression. The other lines in each case are the graphs of the equations corresponding to the original problem's constraint inequalities. Here are the inequalities.

$P + 0.7I \leq 110$	(for the amount of cookie dough)
$0.4I \leq 32$	(for the amount of icing)
$P + I \leq 140$	(for the amount of oven space)
$0.1P + 0.15I \leq 15$	(for the amount of the Woos' preparation time)

Continued on next page

1. Suppose everything is the same as in the original problem, *except* that the Woos have *an unlimited amount of cookie dough.* What combination of plain and iced cookies should the Woos make to maximize their profit?

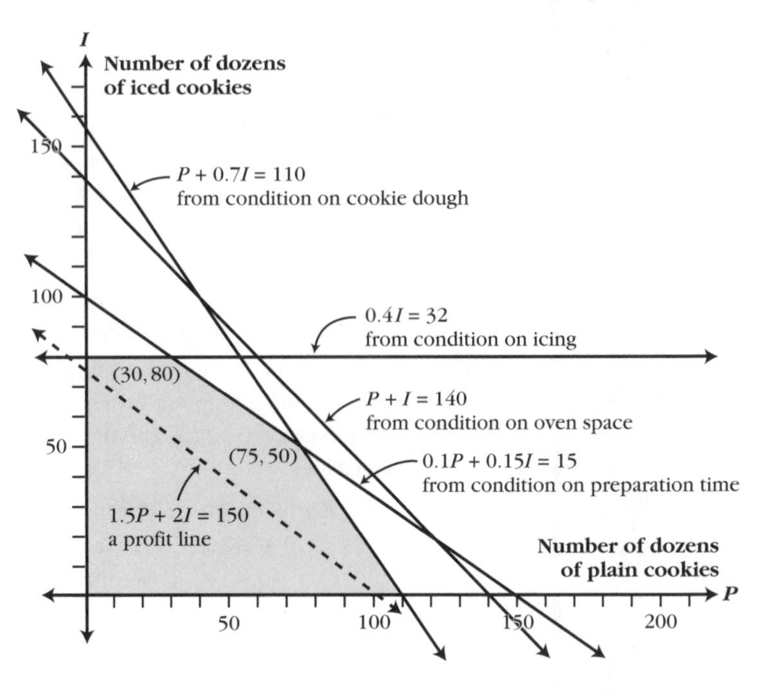

Continued on next page

2. Suppose everything is the same as in the original problem, *except* that the Woos have the additional constraint that they can't sell more than 60 dozen plain cookies. What combination of plain and iced cookies should the Woos make to maximize their profit?

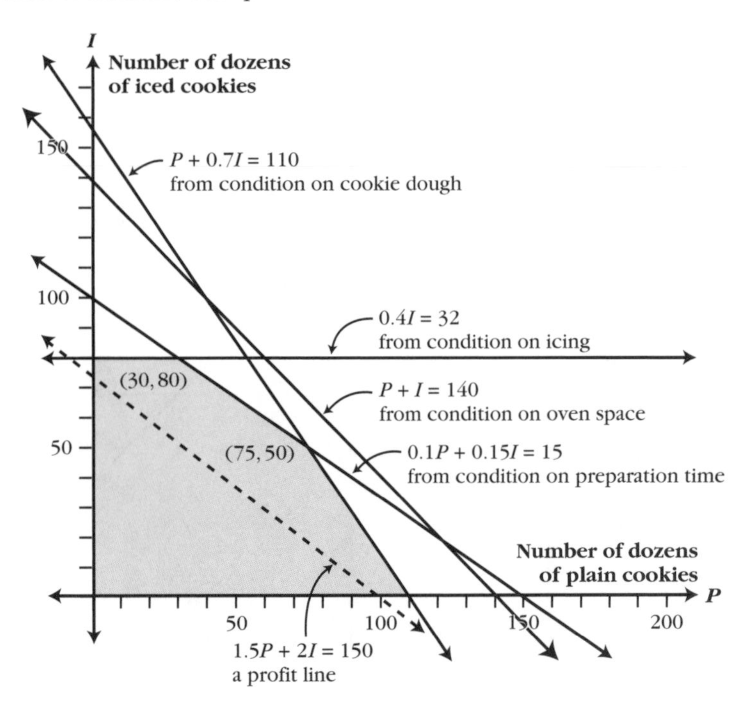

Continued on next page

3. Suppose everything is the same as in the original problem, *except* that the Woos make a profit of \$2.00 on each dozen plain cookies and \$4.00 on each dozen iced cookies (instead of the original profits of \$1.50 and \$2.00). What combination of plain and iced cookies should the Woos make to maximize their profit?

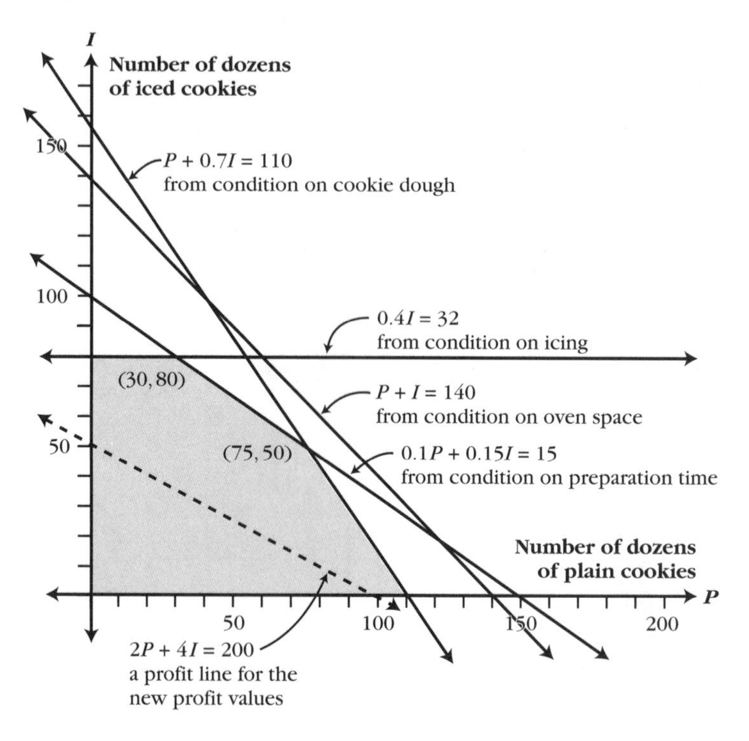

Part II: The Pebbles Rock at Big State U

The Rocking Pebbles are playing a concert at Big State University, where the auditorium seats 2200 people. The concert manager decides to sell some tickets at \$10 each and the rest at \$15 each. How many of each should there be if the manager wants the ticket sales to total \$26,600? (Assume that all the tickets will be sold.)

Find the answer to the problem by setting up and solving a system of two linear equations with two unknowns. Show and explain your work clearly.

DAY 29 Summing Up

Students sum up what they learned in the unit.

Mathematical Topics

- Summarizing the unit

Outline of the Day

1. Discuss unit assessments
2. Sum up the unit

Note: The assessment discussions and unit summary are presented as if they take place on the day following the assessments, but you may prefer to delay this material until you have looked over students' work on the assessments.

These discussion ideas are included here to remind you that some time should be allotted for this type of discussion.

1. Discussion of Unit Assessments

Ask for volunteers to explain their work on each of the problems. Encourage questions and alternate explanations from other students.

- *In-class assessment*

You might begin by getting a consensus about the solution to the equations in Part I. Then have a volunteer propose values and a picture for Question 2 of Part I. Give the rest of the class an opportunity to offer alternatives.

Then move on to Part II, asking for several approaches with explanations.

- *Take-home assessment*

You can have students volunteer to explain their work on each of the problems. Encourage questions and alternative explanations from other students.

On Part I, presenters should explain how the feasible region changes (for Questions 1 and 2), how they identified graphically the point that maximized profit, and how they found the coordinates of that point. For example, on Question 1, the presenter should show how the feasible region is enlarged by eliminating the cookie dough constraint, probably use the "family of parallel lines" method to see that the maximum occurs where the lines $P + I = 140$ and $0.1P + 0.15I = 15$ intersect, and also show how to find the common solution to this pair of equations.

The discussion of Part II can focus primarily on defining variables and setting up equations.

2 Unit Summary

Let volunteers share their portfolio cover letters as a way to start a summary discussion of the unit. Then let students brainstorm about ideas of what they have learned in this unit. This is a good opportunity to review terminology and to place this unit in a broader mathematics context.

Appendix A Supplemental Problems

The appendix contains a variety of activities that you can use to supplement the regular unit material. These activities fall roughly into two categories.

- Reinforcements, which are intended to increase students' understanding of and comfort with concepts, techniques, and methods that are discussed in class and that are central to the unit
- Extensions, which allow students to explore ideas beyond the basic unit and which sometimes deal with generalizations or abstractions of ideas that are part of the main unit

The supplemental activities are given here and in the student materials in the approximate sequence in which you might use them in the unit. In the student book, they are placed following the regular materials for the unit. The discussion below gives specific recommendations about how each activity might work within the unit.

For more ideas about the use of supplemental activities in the IMP curriculum, see the IMP *Teaching Handbook*.

Find My Region (reinforcement)

You can use this problem after discussing *Homework 6: What's My Inequality?* It provides a light-hearted setting in which to practice finding equations for straight-line graphs and inequalities for half planes.

Algebra Pictures (extension)

You might use this problem after discussing *Picturing Cookies—Part II* on Day 7. It extends ideas of the unit by including nonlinear as well as linear inequalities.

Who Am I? (extension or reinforcement)

This logic problem can follow the discussion of *POW 11: A Hat of a Different Color* on Day 10.

More Cereal Variations (extension)

This activity asks students to create some additional variations to the cereal problems (*Homework 10: You Are What You Eat* and

Homework 11: Changing What You Eat). If time allows, you may want to use this activity in class on Day 12 after the homework discussion.

Rap Is Hot! (reinforcement)

You can assign this variation on *Homework 12: Rock 'n' Rap* anytime after the discussion of that assignment. *Note:* You may want to wait until after students have also used the graphing calculators (on Day 14) to examine the situation from *Homework 12: Rock 'n' Rap*.

How Low Can You Get? (extension)

This problem is appropriate for use after Day 13 or so. Students should have had some experience examining the effect of changing parameters in a problem (as in *Homework 11: Changing What You Eat* and in *A Rock 'n' Rap Variation*).

Kick It Harder! (extension)

Students can work on this problem anytime after concluding the discussion of *POW 12: Kick It!* on Day 19.

Shuttling Variations (extension)

This activity presents two major generalizations of the problem in *POW 13: Shuttling Around.* You may want to assign some or all of this activity as part of the POW itself.

And Then There Were Three (extension)

You can have students work on this problem after *Get the Point* and after they have had experience making up problems for two equations in two unknowns in *Homework 21: Inventing Problems.*

An Age-Old Algebra Problem (extension)

You can use this problem as a follow-up to the previous supplemental problem, *And Then There Were Three.*

This page in the student book introduces the supplemental problems.

Much of this unit concerns the use of graphs to understand real-life problems. The supplemental problems for the unit continue the work with graphs and problem solving, and also follow up on some of the POWs. Here are some examples.

- *Find My Region* and *Algebra Pictures* are playful ways in which to explore graphs of equations and inequalities.
- *Rap Is Hot!* is another variation on the situation from *Homework 12: Rock 'n' Rap.*
- *Who Am I?* and *Kick It Harder!* continue the ideas from *POW 11: A Hat of a Different Color* and *POW 12: Kick It!*

Find My Region

This activity is a game for two people, so the first step is to find a partner.

Setting Up a Feasible Region

Each of you needs to define a feasible region using an inequality.

You should both start with the square region in the first quadrant bounded by the inequalities $x \geq 0, x \leq 10, y \geq 0$, and $y \leq 10$. This is the shaded area shown in the graph at the right.

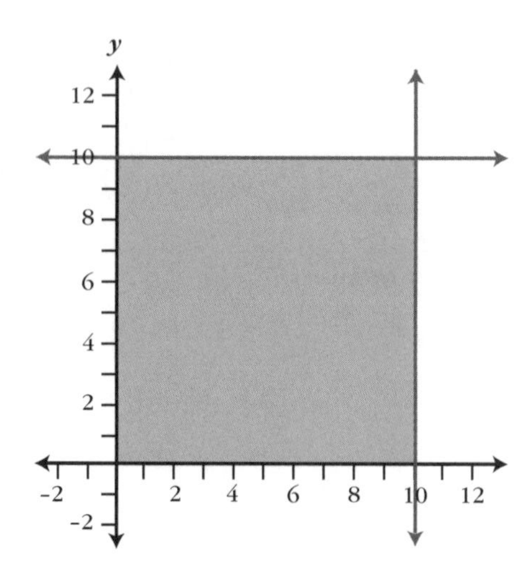

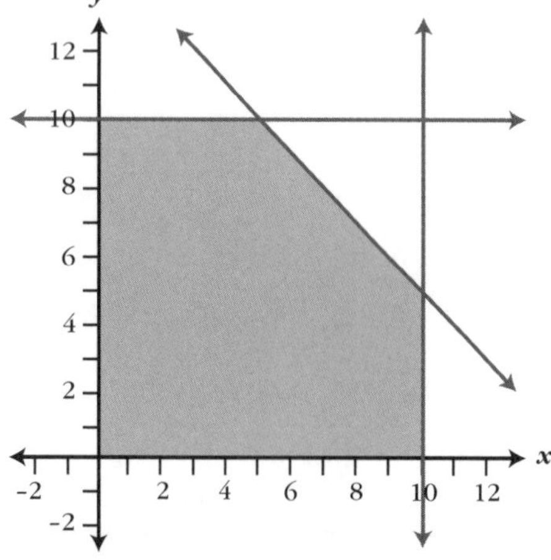

Then each of you needs to choose an inequality to restrict the region further. For example, if you choose the inequality $x + y \leq 15$, your new region will be the shaded area in the graph at the left.

You should sketch your region on a sheet of graph paper.

Continued on next page

Guessing Each Other's Inequalities

Do not tell your partner what your inequality is or show him or her the region you have created. The goal of the game is to figure out what inequality the other player has used.

Sit back to back with your partner so that neither of you can see the other's region. You will each need a blank piece of graph paper to keep track of information you gather about your partner's region.

Here are the rules.

- Take turns guessing a point in your partner's region. For example, you might say, "I guess (3, 6)." (Because you both start with the inequalities $x \geq 0$, $x \leq 10$, $y \geq 0$, and $y \leq 10$, you should only guess points within the square region these inequalities define.)
- Each time one of you guesses a point, the other player will say "inside," "outside," or "boundary," depending on whether the point guessed is inside the region, outside the region, or on one of the boundary lines of the region. For instance, for the region sketched above, (9, 6) would be a boundary point, (5, 7) would be an inside point, and (8, 9) would be an outside point.
- When it is your turn to guess a point, you can choose instead to guess your partner's inequality. For example, you might say, "I guess that $x + y \leq 15$ is your inequality." Your partner must answer truthfully whether your guess is equivalent to his or her inequality. If it is not, it becomes your partner's turn. You do not get a chance to guess a point.
- The winner of the game is the first player to correctly guess the other player's inequality.

Important: When the other player guesses an inequality, you must check whether it is equivalent to the inequality you used. For example, if the other player guesses "$x + y \leq 15$" and you used "$2x + 2y \leq 30$," you must say that this is a correct guess.

Advanced Version

You can make this game more challenging by having each player choose two or three inequalities (in addition to the four inequalities defining the square).

Algebra Pictures

You have been using pictures to help with problems in algebra. In this activity, you will use algebraic inequalities to make artistic pictures (or at least pictures someone might be interested in looking at). But one important change is that these inequalities are no longer all linear.

For each of Questions 1 through 3, make a picture by showing the solution set for the given system of inequalities.

1. $y \leq x + 8$
   $y \leq 16 - x$
   $y \geq (x - 4)^2$

2. $y \leq 2x + 4$
   $y \leq 28 - 2x$
   $y \geq 2$
   $y \leq 10$

3. $y \leq \sqrt{16 - x^2}$
   $y \geq -\sqrt{16 - x^2}$
   $y \leq \frac{1}{2}x^2$

4. Create an interesting picture using systems of inequalities. Your picture can be made up of several parts, such that each individual part is the solution set for a system of inequalities.

Who Am I?

At a college class reunion from dear old Big State U,
I met fifteen classmates, counting men and women, too.
More than half were doctors, and the rest all practiced law.
Of the doctors, more were females, and that I clearly saw.
Even more than female doctors were females doing law,
And these statements all would still be true including me, I saw.
If my friend (a noted lawyer) had a wife and kids at home,
Can you draw any conclusions about me from this poem?

Prove whether you can draw any conclusions about the author from this poem. Show that the line about the friend of the author is needed.

Adapted from *MATHEMATICS: Problem Solving Through Recreational Mathematics,* by B. Averbach and O. Chein, Freeman, 1980, p. 93.

More Cereal Variations

You've seen some problems involving the Hernandez twins and their breakfast habits. In this activity, your task is to make up some variations of your own.

1. Make up a variation on the situation in which the twins would choose to eat just Sweetums.
2. Make up a variation on the situation for which there would be no solution.

Rap Is Hot!

Well, the distributor for Hits on a Shoestring has changed her mind about rap. That is, she has come to believe that rap is more popular in her territory than rock. She now tells the company that it can make up to twice as many rap CDs as rock CDs.

The rest of the facts are the same as in the original *Homework 12: Rock 'n' Rap* problem. Here is a summary of those constraints.

- It costs an average of $15,000 to produce a rock CD and an average of $12,000 to produce a rap CD.
- It takes about 18 hours to produce a rock CD, and about 25 hours to produce a rap CD.
- Hits on a Shoestring must use at least 175 hours of production time.
- Hits on a Shoestring can spend up to $150,000 on production costs next month.
- Each rock CD makes a profit of $20,000 and each rap CD makes a profit of $30,000.

Find out how many CDs of each type Hits on a Shoestring should make next month to maximize its profits, and justify your reasoning. (*Remember:* The company can plan to make a fraction of a CD next month and finish it the month after.)

How Low Can You Get?

Anya and Jesse were working together on a problem, trying to find which point in a certain feasible region minimized various linear expressions involving x and y. Here are the constraints they had.

$y \geq 3$

$y \leq 3x$

$y \leq \frac{1}{2}x + 5$

$y \leq 23 - 4x$

The feasible region for this set of constraints is shown in the graph below.

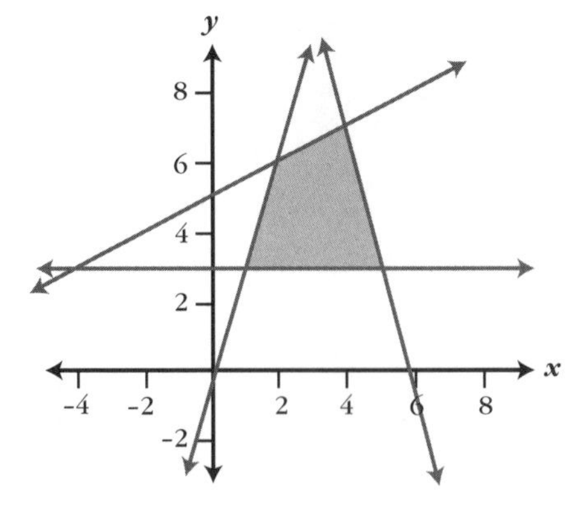

1. Invent two linear expressions in terms of x and y that would have different points in the region as their minimum. Give a convincing argument that the two points really are the minimums for each expression in that region.
2. Invent two different linear expressions that would have the same point in the region as their minimum. Give a convincing argument that the point really is the minimum of each expression in that region.
3. Anya claims that one can pick any point in the region and find a linear expression that has its minimum for the region at that point. Jesse thinks that only certain points can be used as the minimum for a linear expression.

   Who do you think is right? If you agree with Anya, show that any point can be used as a minimum. If you agree with Jesse, explain which points can be used as a minimum and why others cannot.

Kick It Harder!

In *POW 12: Kick It!* you probably found that for some scoring systems, there is a highest impossible score, and for others, there is not. In this activity, your task is to explore this issue further.

1. Find a rule that tells you which scoring systems have no highest impossible score.
2. Find a rule that tells you what the highest impossible score is when there is one.
3. Write a proof of why there is no highest impossible score for those systems for which there is none.

Shuttling Variations

In *POW 13: Shuttling Around,* you are asked to examine a family of puzzles involving the interchange of two sets of markers. In the POW, the number of plain markers is equal to the number of shaded markers, and there is exactly one empty square in the middle. In this activity, you will consider variations on this family of puzzles.

What if the number of markers of each type can be different? For instance, you might consider the following initial situation, with three plain markers and four shaded markers.

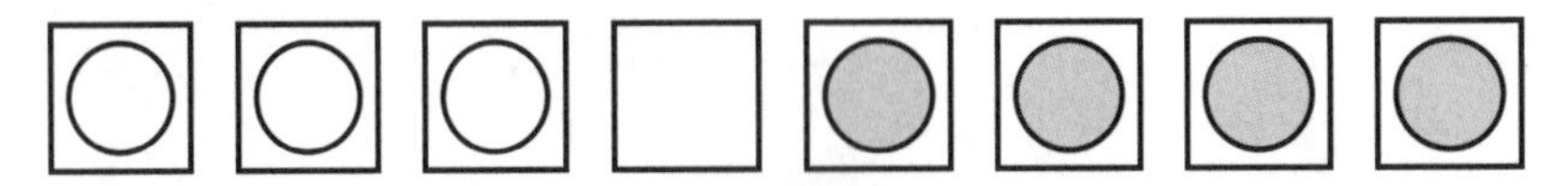

Also, what if there is more than one empty square? For instance, you might consider the following initial situation, with two plain markers, three shaded markers, and two empty squares.

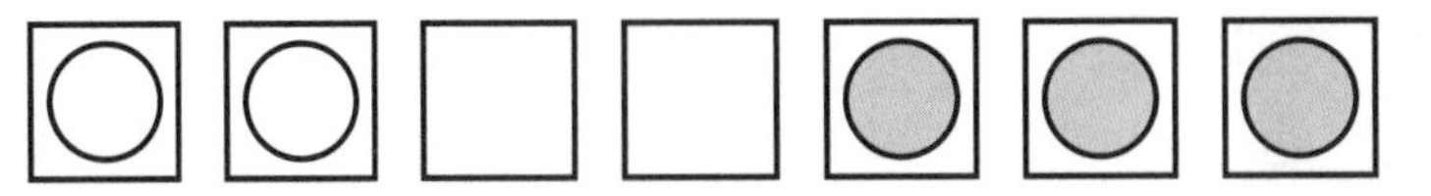

As in the POW, the task is to move the markers so that the shaded markers end up at the left and the plain markers end up at the right, according to these rules.

- The plain markers move only to the right and the shaded markers move only to the left.
- A marker can move to an adjacent open square.
- A marker can jump over *one* marker of the other type into an open square.
- No other types of moves are permitted.

As with the POW itself, you should consider these aspects of the puzzles.

- Determine whether all such puzzles have solutions.
- Look for a rule that describes the minimum number of moves in terms of the number of markers of each type and the number of empty squares.
- Prove your results.

And Then There Were Three

You have done a lot of work in *Cookies* with two linear equations in two unknowns. In this activity, you will extend the ideas and techniques you learned to systems of three linear equations in three unknowns.

1. Here is a problem to solve with three equations in three unknowns.

   I have some dimes, nickels, and quarters. There are 18 coins in all. The total number of dimes and nickels is equal to the number of quarters. The value of my coins is $2.85.

   How many coins of each kind do I have?

   a. Define your variables carefully.

   b. Write three linear equations that express the facts in the situation.

   c. Solve your system of equations.

2. Make up problems for two other situations that can be solved using three variables and three linear equations.

3. Make up two more systems of three linear equations in three unknowns and try to solve them. (You do not need to make up problem situations for these systems.)

4. Write down general directions for finding the common solution for a system of three linear equations in three unknowns.

An Age-Old Algebra Problem

Consider this problem:

> Bob, Maria, and Shoshana all have birthdays on the same day. Bob's present age is two years less than the sum of Shoshana's and Maria's present ages.
>
> In five years, Bob will be twice as old as Maria will be then. Two years ago, Maria was half as old as Shoshana was.
>
> How old is each of them?

1. Define appropriate variables and set up a system of linear equations for this problem. Be especially careful in this problem about how you define your variables.
2. Once you have written the equations, solve them using algebra.

Adapted from *MATHEMATICS: Problem Solving Through Recreational Mathematics,* by B. Averbach and O. Chein, Freeman, 1980, p. 72.

Appendix B Blackline Masters

This appendix contains these items for your use:

- A copy of the feasible region for *Homework 7: Picturing Pictures*, for making a transparency
- A form for students to use in evaluating one another's presentations for the activity *Producing Programming Problems*
- In-class and take-home unit assessments for *Cookies,* which you will need to reproduce for students for Day 28

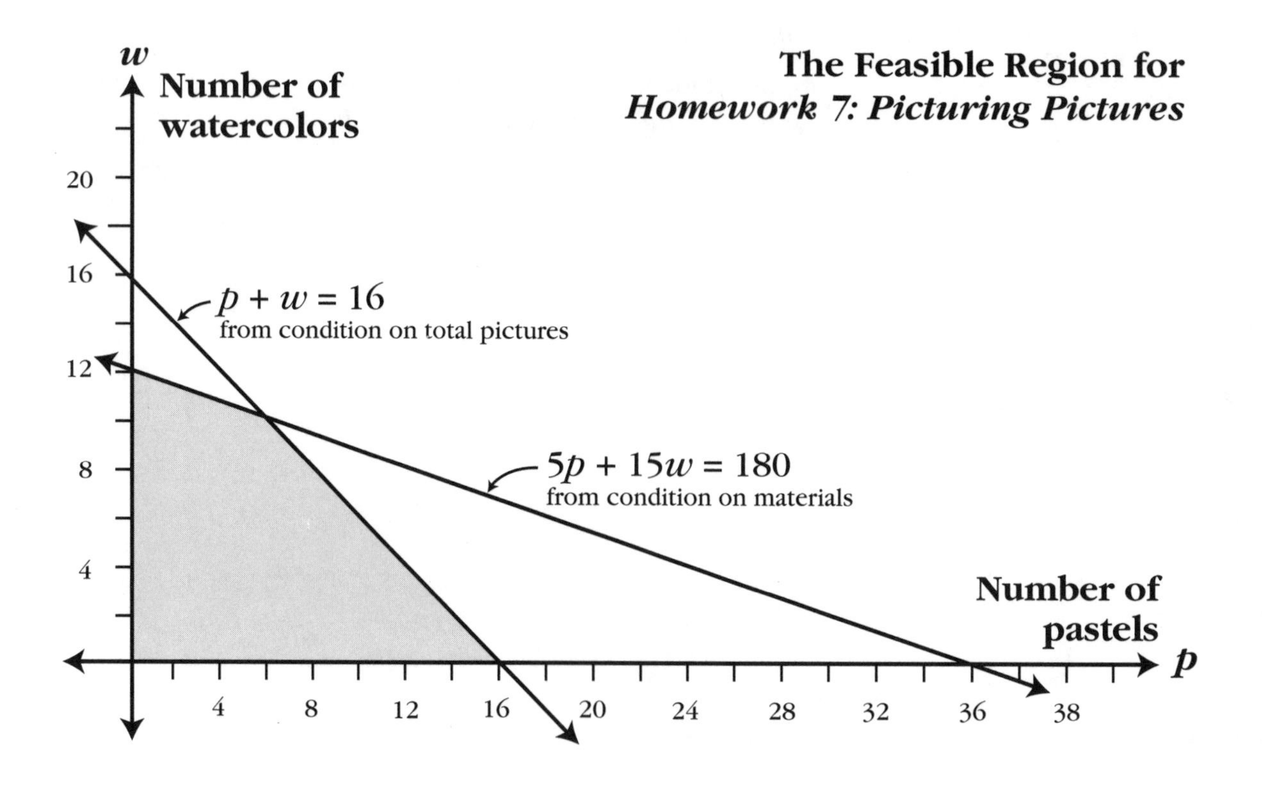

w
Number of watercolors
The Feasible Region for
Homework 7: Picturing Pictures
20
16
12
8
4
p + w = 16
from condition on total pictures
5p + 15w = 180
from condition on materials
Number of pastels
p
4
8
12
16
20
24
28
32
36
38

Grading Programming Problem Presentations

Your group has the responsibility for assigning a letter grade to your fellow students on their presentation. Use the criteria established by the class as the basis for the grade you give.

Presenting group's group number: ______________________________

Your group number: ______________________________

Grade you are giving: ______________________________

Reasons for this grade: ______________________________

____

____

____

____

____

____

____

____

In-Class Assessment for "Cookies"

Part I: Graph It

Consider the following system of two equations:

$$5x + 7y = 200$$
$$9x - 3y = 85$$

1. Use your graphing calculator to estimate the solution to the system.

2. Set your viewing window so that the coordinate axes, the graphs of both equations, and the point of intersection of the two graphs are all visible.

   a. Give the values below to describe your viewing window.
      - Minimum value of x
      - Maximum value of x
      - Minimum value of y
      - Maximum value of y

   b. Make a sketch of what you see on your graphing calculator screen, and give the approximate coordinates of the solution.

Part II: Solve It

Use algebra to solve each system of equations. Show and explain your work clearly.

3. $4x + 3y = 5$
   $2x - 5y = 9$

4. $4x - 6y = 20$
   $6x - 9y = 24$

Take-Home Assessment for *Cookies*

Part I: What If . . . ?

The problems in this part of the assessment are variations on the unit problem, *How Many of Each Kind?* In each problem, you should find the combination of plain and iced cookies that maximizes the Woos' profit for the new situation, and then explain your answer. Consider these three variations from the original situation as *three separate problems.*

Each problem is accompanied by a graph that shows the feasible region of the original problem, with the shaded area representing that original feasible region.

Questions 1 and 2 show a profit line based on the original problem. Question 3 shows a profit line based on a different profit expression. The other lines in each case are the graphs of the equations corresponding to the original problem's constraint inequalities. Here are the inequalities.

$P + 0.7I \leq 110$ (for the amount of cookie dough)

$0.4I \leq 32$ (for the amount of icing)

$P + I \leq 140$ (for the amount of oven space)

$0.1P + 0.15I \leq 15$ (for the amount of the Woos' preparation time)

Continued on next page

1. Suppose everything is the same as in the original problem, *except* that the Woos have *an unlimited amount of cookie dough.* What combination of plain and iced cookies should the Woos make to maximize their profit?

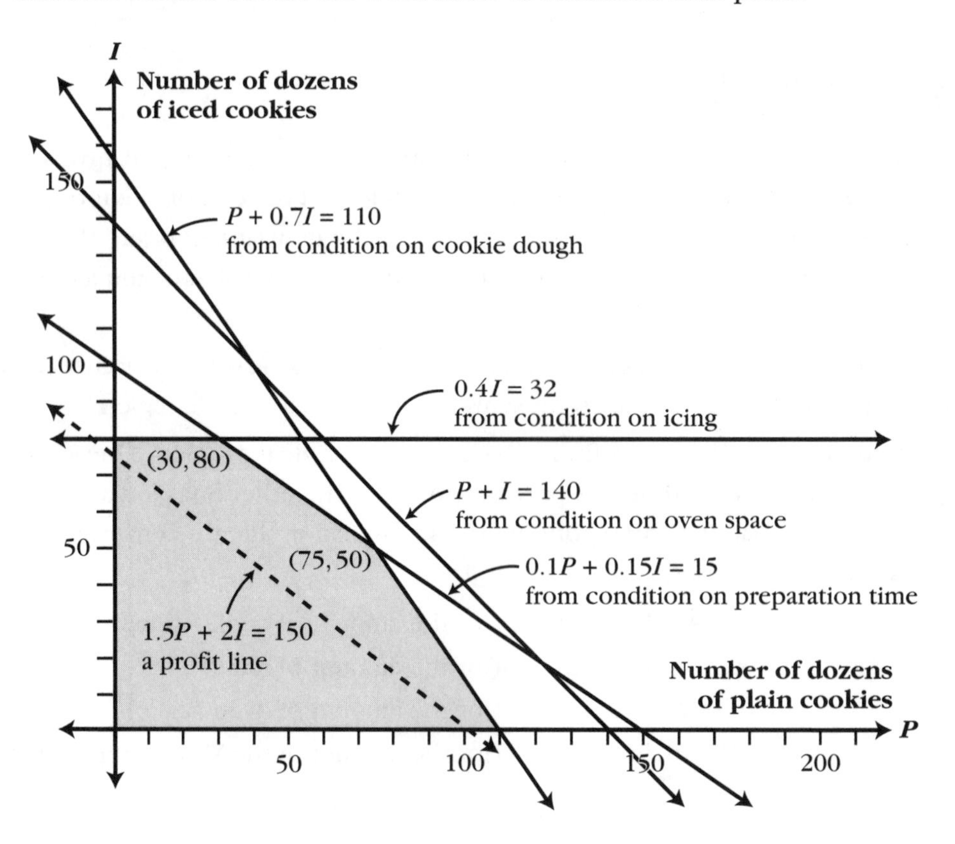

Continued on next page

2. Suppose everything is the same as in the original problem, *except* that the Woos have the additional constraint that they can't sell more than 60 dozen plain cookies. What combination of plain and iced cookies should the Woos make to maximize their profit?

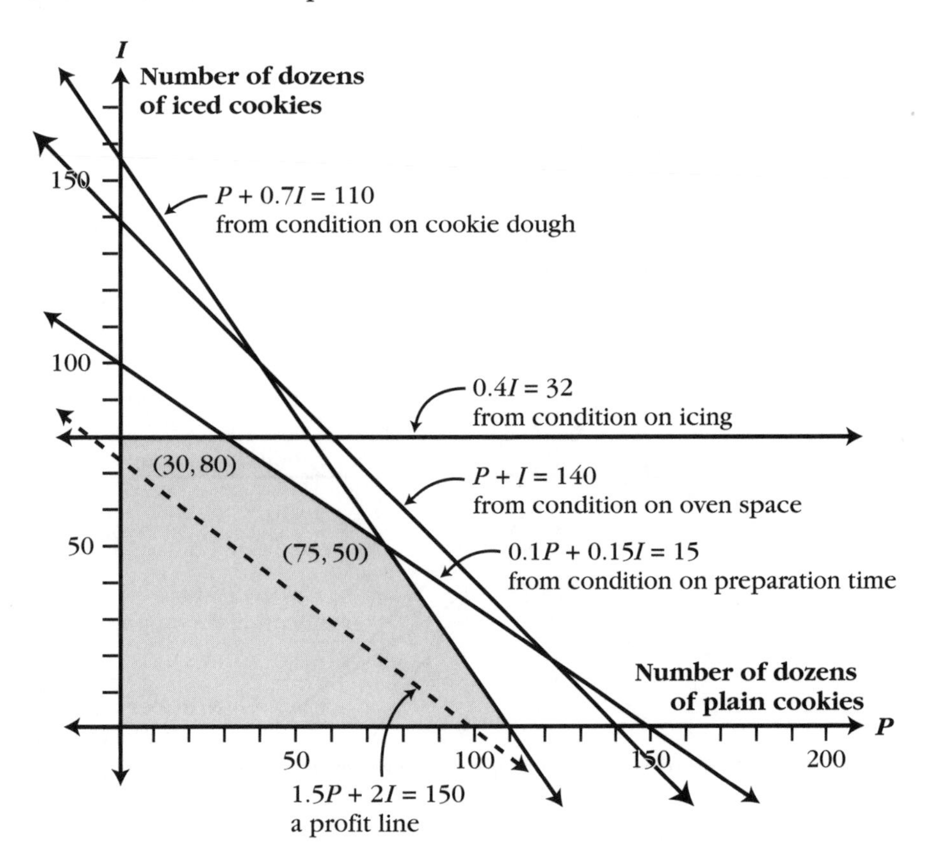

Continued on next page

3. Suppose everything is the same as in the original problem, *except* that the Woos make a profit of \$2.00 on each dozen plain cookies and \$4.00 on each dozen iced cookies (instead of the original profits of \$1.50 and \$2.00). What combination of plain and iced cookies should the Woos make to maximize their profit?

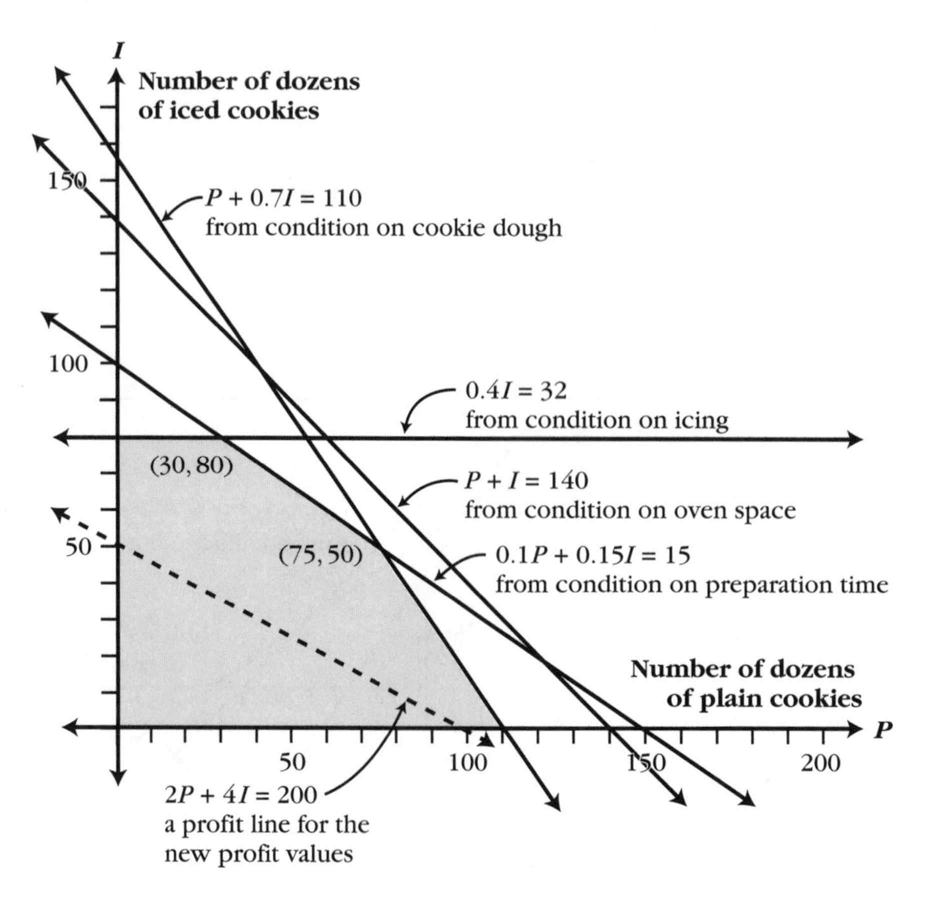

Part II: The Pebbles Rock at Big State U

The Rocking Pebbles are playing a concert at Big State University, where the auditorium seats 2200 people. The concert manager decides to sell some tickets at \$10 each and the rest at \$15 each. How many of each should there be if the manager wants the ticket sales to total \$26,600? (Assume that all the tickets will be sold.)

Find the answer to the problem by setting up and solving a system of two linear equations with two unknowns. Show and explain your work clearly.

This is the glossary for all five units of IMP Year 2.

Absolute growth The growth of a quantity, usually over time, found by subtracting the initial value from the final value. Used in distinction from **percentage growth.**

Additive law of exponents The mathematical principle which states that the equation

$$A^B \cdot A^C = A^{B+C}$$

holds true for all numbers A, B, and C (as long as the expressions are defined).

Altitude of a parallelogram or trapezoid A line segment connecting two parallel sides of the figure and perpendicular to these two sides. Also, the length of such a line segment. Each of the two parallel sides is called a **base** of the figure.

Examples: Segment $\overline{KL}$ is an altitude of parallelogram $GHIJ$, with bases $\overline{GJ}$ and $\overline{HI}$ and segment $\overline{VW}$ is an altitude of trapezoid $RSTU$, with bases $\overline{RU}$ and $\overline{ST}$.

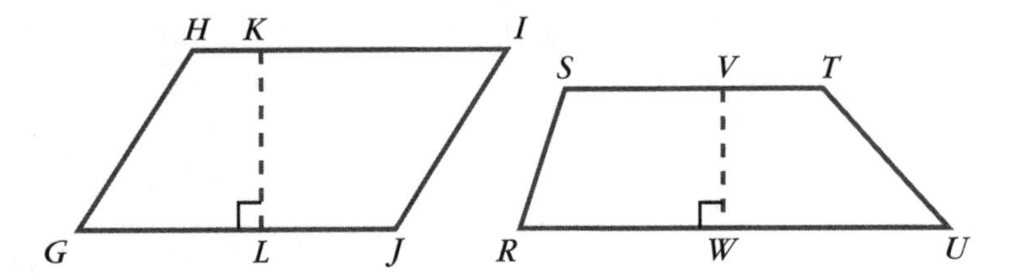

Altitude of a triangle

A line segment from any of the three vertices of a triangle, perpendicular to the opposite side or to an extension of that side. Also, the length of such a line segment. The side to which the perpendicular segment is drawn is called the **base** of the triangle and is often placed horizontally.

Example: Segment $\overline{AD}$ is an altitude of triangle ABC. Side BC is the base corresponding to this altitude.

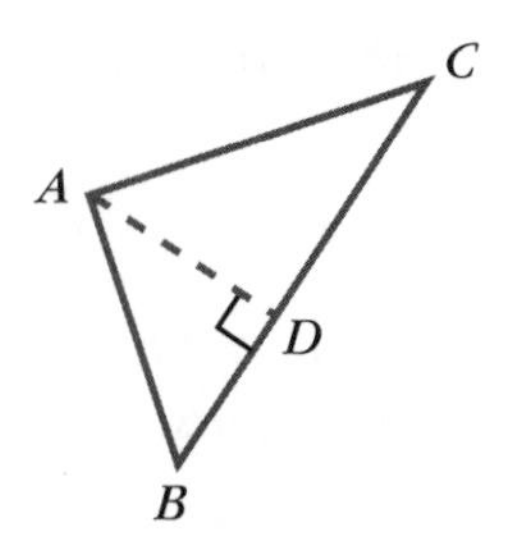

Base

The side of a triangle, a parallelogram, or a trapezoid to which an altitude is drawn. For **base of a prism,** see *The World of Prisms* in the unit *Do Bees Build It Best?*

Chi-square statistic

A number used for evaluating the statistical significance of the difference between observed data and the data that would be expected under a specific hypothesis. The chi-square (χ^2) statistic is defined as a sum of terms of the form

$$\frac{(\text{observed} - \text{expected})^2}{\text{expected}}$$

with one term for each observed value.

Composite number

A counting number having more than two whole-number divisors.

Cosecant

See *A Trigonometric Summary* in *Do Bees Build It Best?*

Cosine

See *A Trigonometric Summary* in *Do Bees Build It Best?*

Cotangent

See *A Trigonometric Summary* in *Do Bees Build It Best?*

Dependent equations

See **system of equations.**

Distributive property — The mathematical principle which states that the equation $a(b + c) = ab + ac$ holds true for all numbers a, b, and c.

Edge — See **polyhedron.**

Equivalent equations (or inequalities) — A pair of equations (or inequalities) that have the same set of solutions.

Equivalent expressions — Algebraic expressions that give the same numerical value no matter what values are substituted for the variables.

Example: $3n + 6$ and $3(n + 2)$ are equivalent expressions.

Expected number — The value that would be expected for a particular data item if the situation perfectly fit the probabilities associated with a given hypothesis.

Face — See **polyhedron.**

Factoring — The process of writing a number or an algebraic expression as a product.

Example: The expression $4x^2 + 12x$ can be factored as the product $4x(x + 3)$.

Feasible region — The region consisting of all points whose coordinates satisfy a given set of constraints. A point in this set is called a **feasible point.**

Geometric sequence — A sequence of numbers in which each term is a fixed multiple of the previous term.

Example: The sequence 2, 6, 18, 54, . . . , in which each term is 3 times the previous term, is a geometric sequence.

Hypothesis — Informally, a theory about a situation or about how a certain set of data is behaving. Also, a set of assumptions used to analyze or understand a situation.

Hypothesis testing — The process of evaluating whether a hypothesis holds true for a given population. Hypothesis testing usually involves statistical analysis of data collected from a sample.

Inconsistent equations — See **system of equations.**

Independent equations — See **system of equations.**

Inverse trigonometric function — Any of six functions used to determine an angle if the value of a trigonometric function is known.

Example: For x between 0 and 1, the inverse sine of x (written $\sin^{-1}x$) is defined to be the angle between 0° and 90° whose sine is x.

Lateral edge or face — See *The World of Prisms* in *Do Bees Build It Best?*

Lateral surface area — See *The World of Prisms* in *Do Bees Build It Best?*

Law of repeated exponentiation — The mathematical principle which states that the equation

$$\left(A^B\right)^C = A^{BC}$$

holds true for all numbers A, B, and C (as long as the expressions are defined).

Linear equation — For two variables, an equation whose graph is a straight line. More generally, an equation stating that two linear expressions are equal.

Linear expression — For a single variable x, an expression of the form $ax + b$, where a and b are any two numbers, or any expression equivalent to an expression of this form. For more than one variable, any sum of linear expressions in those variables (or an expression equivalent to such a sum).

Example: $4x - 5$ is a linear expression in one variable; $3a - 2b + 7$ is a linear expression in two variables.

Linear function — For functions of one variable, a function whose graph is a straight line. More generally, a function defined by a linear expression.

Example: The function g defined by the equation $g(t) = 5t + 3$ is a linear function in one variable.

Linear inequality
An inequality in which both sides of the relation are linear expressions.

Example: The inequality $2x + 3y < 5y - x + 2$ is a linear inequality.

Linear programming
A problem-solving method that involves maximizing or minimizing a linear expression, subject to a set of constraints that are linear equations or inequalities.

Logarithm
The power to which a given base must be raised to obtain a given numerical value.

Example: The expression $\log_2 28$ represents the solution to the equation $2^x = 28$. Here, "log" is short for *logarithm,* and the whole expression is read "log, base 2, of 28."

Net
A two-dimensional figure that can be folded to create a three-dimensional figure.

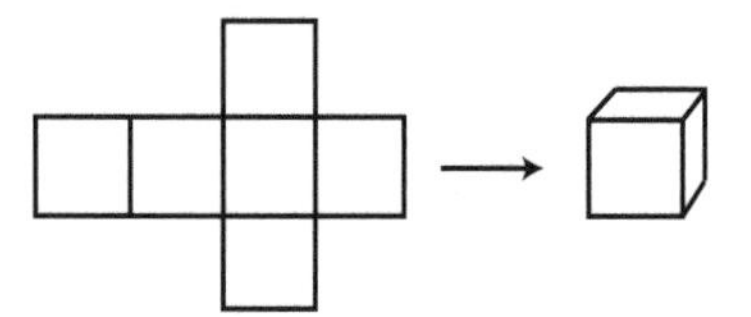

Example: The figure on the left is a net for the cube.

Normal distribution
See *Normal Distribution and Standard Deviation Facts* in *Is There Really a Difference?*

Null hypothesis
A "neutral" assumption of the type that researchers often adopt before collecting data for a given situation. The null hypothesis often states that there are no differences between two populations with regard to a given characteristic.

Order of magnitude
An estimate of the size of a number based on the value of the exponent of 10 when the number is expressed in scientific notation.

Example: The number 583 is of the second order of magnitude because it is written in scientific notation as $5.83 \cdot 10^2$, using 2 as the exponent for the base 10.

Parallelogram A quadrilateral in which both pairs of opposite sides are parallel.

Example: Polygons *ABCD* and *EFGH* are parallelograms.

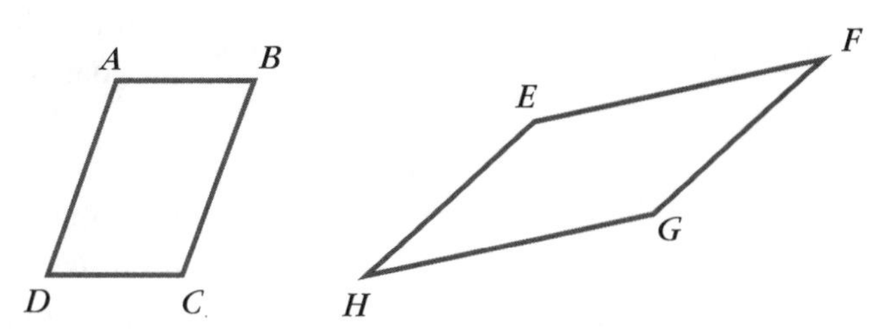

Percentage growth The proportional rate of increase of a quantity, usually over time, found by dividing the absolute growth in the quantity by the initial value of the quantity. Used in distinction from **absolute growth.**

Polygon A closed two-dimensional figure consisting of three or more line segments. The line segments that form a polygon are called its sides. The endpoints of these segments are called **vertices** (singular: **vertex**).

Examples: All the figures below are polygons.

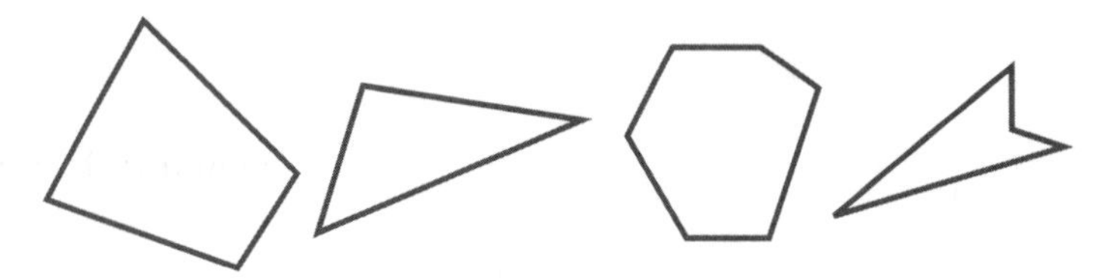

Polyhedron A three-dimensional figure bounded by intersecting planes. The polygonal regions formed by the intersecting planes are called the **faces** of the polyhedron, and the sides of these polygons are called the **edges** of the polyhedron. The points that are the vertices of the polygons are also **vertices** of the polyhedron.

Example: The figure below shows a polyhedron. Polygon *ABFG* is one of its faces, segment $\overline{CD}$ is one of its edges, and point *E* is one of its vertices.

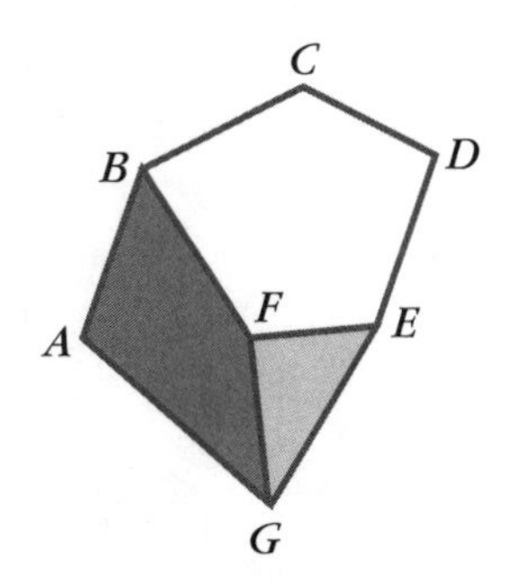

Population — A set (not necessarily of people) involved in a statistical study and from which a sample is drawn.

Prime factorization — The expression of a whole number as a product of prime factors. If exponents are used to indicate how often each prime is used, the result is called the **prime power factorization.**

Example: The prime factorization for 18 is $2 \cdot 3 \cdot 3$. The prime power factorization for 18 is $2^1 \cdot 3^2$.

Prime number — A counting number that has exactly two whole-number divisors, 1 and itself.

Prism — A type of polyhedron in which two of the faces are parallel and congruent. For details and related terminology, see *The World of Prisms* in *Do Bees Build It Best?*

Profit line — In the graph used for a linear programming problem, a line representing the number pairs that give a particular profit.

Pythagorean theorem — The principle for right triangles which states that the sum of the squares of the lengths of the two legs equals the square of length of the hypotenuse.

Example: In right triangle *ABC* with legs of lengths a and b and hypotenuse of length c, the Pythagorean theorem states that $a^2 + b^2 = c^2$.

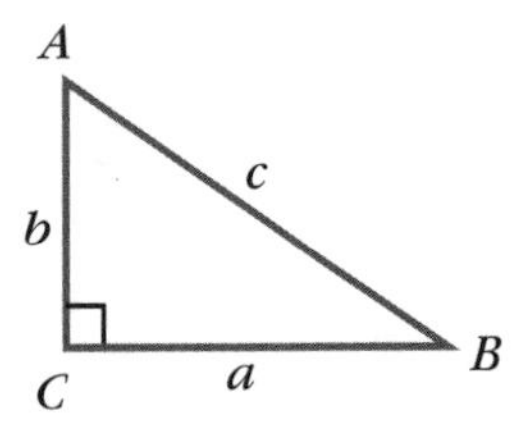

Right rectangular prism — See *The World of Prisms* in *Do Bees Build It Best?*

Sample — A selection taken from a population, often used to make conjectures about the entire population.

Sampling fluctuation — Variations in data for different samples from a given population that occur as a natural part of the sampling process.

Scientific notation — A method of writing a number as the product of a number between 1 and 10 and a power of 10.

Example: The number 3158 is written in scientific notation as $3.158 \cdot 10^3$.

Secant — See *A Trigonometric Summary* in *Do Bees Build It Best?*

Sine — See *A Trigonometric Summary* in *Do Bees Build It Best?*

Standard deviation — See *Normal Distribution and Standard Deviation Facts* in *Is There Really a Difference?*

Surface area — The amount of area that the surfaces of a three-dimensional figure contain.

System of equations — A set of two or more equations being considered together. If the equations have no common solution, the system is **inconsistent.** Also, if one of the equations can be removed from the system without changing the set of common solutions, that equation is **dependent** on the others, and the system as a whole is also **dependent.** If no equation is dependent on the rest, the system is **independent.**

In the case of a system of two linear equations with two variables, the system is *inconsistent* if the graphs of the two equations are distinct parallel lines, *dependent* if the graphs are the same line, and *independent* if the graphs are lines that intersect in a single point.

Tangent — See *A Trigonometric Summary* in *Do Bees Build It Best?*

Tessellation — Often, a pattern of identical shapes that fit together without overlapping.

Trapezoid A quadrilateral in which one pair of opposite sides is parallel and the other pair is not.

Example: Polygons *KLMN* and *PQRS* are trapezoids.

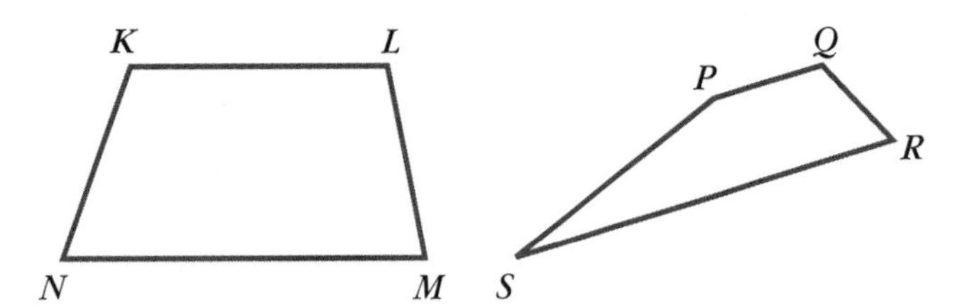

Trigonometry For a right triangle, the study of the relationships between the acute angles of the triangle and the lengths of the sides of the triangle. For details, see *A Trigonometric Summary* in the unit *Do Bees Build It Best?*

Vertex See **polygon** and **polyhedron.**

x-intercept A point where a graph crosses the x-axis. Sometimes, the x-coordinate of that point.

y-intercept A point where a graph crosses the y-axis. Sometimes, the y-coordinate of that point.

Photographic Credits

Teacher Book Classroom Photography

14 Sharon G. Taylor, Damien Mancha; **48** Hillary Turner; **71** Pleasant Valley High School, Mike Christensen; **110** Hillary Turner; **125** Sharon G. Taylor, Damien Mancha; **138** Sharon G. Taylor, Damien Mancha; **167** Sharon G. Taylor, Damien Mancha; **186** Sharon G. Taylor, Damien Mancha

Student Book Interior Photography

304 Hillary Turner; **311** Colton High School, Sharon Taylor; **325** Pleasant Valley High School, Mike Christensen; **335** Comstock © 1995; **336** Comstock © 1996; **339** Pleasant Valley High School, Mike Christensen; **347** Cappucino High School, Chicha Lynch; **354** Pleasant Valley High School, Mike Christensen; **370** Comstock © 1996

Cover Photography and Cover Illustration

Background © Tony Stone Worldwide **Top left to bottom right** from *Alice in Wonderland* by Lewis Carroll; Hillary Turner; Hillary Turner; © Image Bank

Front Cover Students

Colin Bjorklund, Liana Steinmetz, Sita Davis, Thea Singleton, Jenée Desmond, Jennifer Lynn Anker, Lidia Murillo, Keenzia Budd, Noel Sanchez, Seogwon Lee, Kolin Bonet (photographed by Hillary Turner)